ELMHURST PUBLIC LIBRARY
3 1135 01764 7253

Cells to Civilizations

Cells to Civilizations

The Principles of Change That Shape Life

Enrico Coen

PRINCETON UNIVERSITY PRESS
PRINCETON AND OXFORD

Published by Princeton University Press,
41 William Street, Princeton, New Jersey 08540
In the United Kingdom: Princeton University Press,
6 Oxford Street, Woodstock, Oxfordshire OX20 1TW
press.princeton.edu

Jacket illustration of coral is courtesy of *Haeckel's Art Forms from Nature*.
Jacket photographs: "Old Wooden Wheel" © Hisom Silviu; "Green Gecko on the Roof" (Zurich zoo) © Natali Glado. Photographs are courtesy of Shutterstock.

Library of Congress Cataloging-in-Publication Data
Coen, Enrico.
Cells to civilizations : principles of change that shape life / Enrico Coen.
p. cm.
Includes bibliographical references and index.
ISBN 978-0-691-14967-7 (cloth : alk. paper) 1. Life (Biology) 2. Evolution (Biology) 3. Developmental biology. 4. Learning. 5. Social change. I. Title.
QH501.C63 2012
576.8—dc23 2011045729

British Library Cataloging-in-Publication Data is available
This book has been composed in Perpetua Std
Book design by Marcella Engel Roberts
Printed on acid-free paper. ∞
Printed in the United States of America

10 9 8 7 6 5 4 3 2 1

Contents

INTRODUCTION • A RECIPE FOR CHANGE 1

History and Form 3

Life's Creative Recipe 8

ONE • LOOPS AND LOTTERIES 13

Principle of Population Variation 16

Principle of Persistence 20

Principle of Reinforcement 23

Principle of Competition 24

Combining Principles 29

TWO • FROM GENES TO ECOSYSTEMS 34

Principle of Cooperation 36

Principle of Combinatorial Richness 40

Wandering Clouds 44

Principle of Recurrence 48

The Origin of Species 52

Species and Ecosystems 57

A Recipe for Evolution 58

THREE • CONVERSATIONS OF AN EMBRYO 61

Turing's Principles 63

Patterning a Cell 68

Switching Genes On and Off 70

A Molecular Fight 73

Looking into Gradients 76

A Common Form 80

FOUR • COMPLETING THE PICTURE 84

An Embryonic Cocktail Party 85

A Cooperative Effort 87

Regulatory Riches 88

Building on the Past 90

The Expanding Canvas 93
Deformation 97
The Three-Dimensional Canvas 105
A Common Recipe 108

FIVE • HISTORY IN THE MAKING 111
Unicellular Beginnings 114
Moving up a Scale 115
Zooming and Growing 118
A Recipe within a Recipe 120

SIX • HUMBLE RESPONSES 122
Making Adjustments 124
Flora's Story 127
The Bite of Venus 132
The Sensible Sea Slug 135
Patterns in Time 140
Human Responses 143
Carving up the World 147

SEVEN • THE NEURAL SIBYL 150
The Prophetic Dog 152
Predictive Neurons 155
Learning from Discrepancies 157
Pavlov and Punishments 163
Core Principles 164
A Neural Journey 168
Staying on the Move 169
A Recipe for Learning 173

EIGHT • LEARNING THROUGH ACTION 175
Calibration 176
Jumping Eyes 178
Visual Shifts 181
Learning to Calibrate 184
Action-Learning Loops 187
Smooth Movements 188
An Active Journey 193
Learning with Others 197

NINE • SEEING AS 200
The Neural Eye 204
Multiple Eyes 208
Seeing through Models 211
Learning at Many Levels 213
Top-down and Bottom-up 216
Competing Interpretations 217
A Question of Style 220
Creative Acts 228

TEN • FRAMING RECIPES 232
Development of Learning 233
Basic Instincts 237
Flexibility versus Directness 242

ELEVEN • THE CRUCIBLE OF CULTURE 244
The Apprentice 245
Fruitful Populations 248
Lasting Change 250
Cultural Reinforcement 252
The Force of Competition 254
Cooperative Efforts 255
A Cultural Mix 257
Propelled by the Past 260
A Cultural Recipe 263

TWELVE • THE GRAND CYCLE 266
Cultural Origins 267
Possible Worlds 270
Nature's Self-Portrait 275

Acknowledgments 281
Notes 283
References 299
Illustration Credits 307
Index 313

Cells to Civilizations

INTRODUCTION
A Recipe for Change

LIFE HAS REMARKABLE transforming powers. Through billions of years of evolution, elementary forms are transformed into the complex creatures of today. Over nine months a nondescript egg turns itself into a human being. During a few years a flailing baby becomes a walking, talking, and reasoning adult. And over ten thousand years, human societies are transformed from small tribal communities into the complex cities and civilizations of today.

It is tempting to think that this transformative property of life depends on a single underlying mechanism. Yet when we look, we see four very different mechanisms at play. All the creatures on our planet are thought to have arisen as a result of individuals competing for survival and reproduction over many generations, through the process of Darwinian selection. Quite a different mechanism underlies the development of an egg. As a fertilized egg goes through repeated rounds of division, patterns are built up in the embryo by cells signaling and responding to each other in various ways. Development is about patterning within a growing embryo rather than competition for reproductive success. Yet another mechanism underlies learning. As an animal interacts with its surroundings, neural connections in the brain are modified. Some connections are lost or weakened while others are formed or strengthened, allowing new relationships in the environment to be captured. Learning is all about modifying neural interactions and connections. Finally we come to cultural change. Humans interact within social groups, leading to advances in tools, artifacts, and knowledge. Culture is a social phenomenon that depends on how we as humans behave and interact.

There is no obvious commonality among the operations of these various processes. They all seem to happen in completely different ways: evolution by differential reproductive success, development by cell proliferation growth and patterning, learning by changes in neural connections, and cultural change through human interactions.

It seems strange that Nature should have come up with four entirely different transformation mechanisms. Just as physicists strive for a "theory of everything," one that brings together its fundamental theories, so you might expect biologists to be seeking a unified theory for living transformations; a theory that encompasses evolution, development, learning, and cultural change. Some attempts at unification have been made in the past. Ernst Haeckel, a keen follower of Darwin in the nineteenth century, proposed a direct relationship between evolution and development. He thought that as an embryo develops, it reenacts its evolutionary history. Thus, human embryos would pass through a fishy stage, a reptilian stage, and so on, as they grow in the womb. This idea, however, was later shown to be misguided. A more recent attempt at bringing together evolution and learning was Gerald Edelman's theory of "neural Darwinism." In the 1980s Edelman proposed that neural patterns in the brain are selected for during learning in a way that parallels natural selection. This idea, however, has been heavily criticized. It seems that a unified explanation of diverse living transformations is fraught with difficulties.

Perhaps Nature really has four completely different ways of transforming itself and we should look no further. I believe this view is mistaken. In this book I want to show that recent advances in our scientific understanding have given us access to a unified picture of how living systems transform themselves, from the origins of bacteria to the creation of an artistic masterpiece. For the first time, we can begin to see a common set of ingredients and mechanisms underlying life's transformations.

Why should finding common ingredients matter? After all, studies on evolution, development, learning, and culture seem to have so far moved along quite happily without concern for overarching similarities. What do we gain from viewing them collectively? Suppose you compare the way ice melts with how water boils. These processes differ in many ways: the first involves a solid turning into a liquid around zero degrees centigrade, while the second involves a liquid turning into a gas at one hundred degrees. Yet these two transitions can also be seen to have many features in common, for both involve a change in the strength and energy of interactions among water molecules. They are different manifestations of the same underlying process. This unifying perspective gives us a deeper understanding of what is happening than what we perceive by simply viewing each transition in isolation. In the same way, looking at the common elements behind different living transformations can help us to understand the essence of each process, while also giving us a broader overview of events.

Such an approach is perhaps reasonable for evolution, development, and learning as they are all subjects of intense scientific investigation. It might, however, seem a little far-fetched to extend this approach to cultural change. We tend to think of human creativity and culture as being so complicated and special to us that science can have little to say about them. But when it comes to looking at living transformations as a whole, we see that science plays a dual role. On the one hand, science provides a source of knowledge about the world and our place within it, and thus frames our culture. On the other hand, science is a product of culture, the result of humans working collectively over many years to make sense of the world around them. It is only when we look at all types of living transformation together that we can get a clear view of this dual aspect of our outlook, of how science both frames our culture and is framed by it. We then gain not only a broader understanding of how cultural change is brought about, but also a better understanding of how it is related to our biological past.

Why has it taken so long to arrive at this collective viewpoint?

History and Form

On the face of it, war and chess are very different. War involves people fighting and killing each other, while chess has two people peacefully sitting at a board pushing some pieces of wood around. Yet, in spite of such obvious differences, these activities are closely related. First they are connected in history. The origins of chess can be traced back to the game of chatrang played in Persia in the fifth to sixth centuries, which in turn may have derived from the Indian game of chaturanga. Like modern chess, chatrang was a game for two players with thirty-two pieces on a board of sixty-four squares. Each player had an army comprising two elephants, two horses, two chariots, and eight foot soldiers. These were based on the main fighting units of the time and were the predecessors of bishops, knights, rooks, and pawns in modern chess. There was also a king and minister (equivalent to the modern queen). The aim was to capture or trap the opponent's king. The game illustrated how one army could overcome and outflank another through strategy and cunning. Chatrang was as much a war game in its time as computer war games are today.

There are other ways in which chess can be related to war that do not depend on knowing their historical links. Both can be seen to be highly competitive, involving opponents trying to overcome each other. Both are territorial, with each competitor trying to occupy or control regions. They both involve attempts to eliminate or capture elements of the opposition. And there is a strong strategic element of cooperation within each side, with units supporting each other when mounting an attack or maintaining a defense. War and chess have a similar form as well as being connected in history.

These two ways of relating chess to war, through history or form, are themselves related. Similarities in form, such as competition and territoriality, reflect the origins of chess as a game that mirrors war. Not all features of war, however, are mirrored by chess. Elements such as the physical death of humans are not a part of chess; nor is consideration of varying layouts of land or environmental conditions. Chess is always played with the same arrangement of squares and perfectly controlled initial conditions, and notions such as weather or visibility don't enter into the game. Chess is not simply a mirror of war; it is an abstraction of war that captures a particular set of elements. It is these essential features that provide the similarities in form.

The various transformations of life may also be related through history or form. It is thought that life on Earth originated about 3.8 billion years ago, and by 3.5 billion years ago our planet was populated by a diverse collection of single-celled creatures. At this stage of evolution there were no complex many-celled organisms. These evolved later, during the last billion years or so, through the process of development from fertilized eggs. Development, the transformation of eggs into multicellular organisms, arose long after evolution started, just as the game of chess arose many years after wars had begun to be fought.

Similarly, learning arose after development. The first multicellular creatures to develop on our planet were hardly able to learn. They did not have brains that allowed new relationships in their surroundings to be captured. Complex nervous systems started to evolve later. This happened through changes in the way embryos developed. Part of the development of some creatures became dedicated to the formation of brains with connecting nerve pathways. With that came the ability of an organism to learn from its surroundings, an ability that is now shared by many of our cousins, from slugs and dogs to chimpanzees. Learning came after development, just as development came after evolution.

The final type of living transformation to arise was cultural change. As humans spread over the earth in social groups, learning how to domesticate and exploit other creatures for their own benefit, they generated a food surplus. This excess, together with the human capacity to learn and innovate, allowed societies to support and develop a range of human specializations, such as builders, soldiers, artists, teachers, and administrators, leading to the formation of elaborate cultural systems. Civilizations only began to emerge within the last ten thousand years, long after the ability to learn first arose. So, cultural change is much more recent than any of the other processes.

The historical relationships seem clear—first came evolution, then development, then learning, and finally cultural change. This time sequence reflects the dependence of each process on its predecessor. The ability of eggs to develop into multicellular creatures arose through the process of evolution. The ability to learn depends on the prior development of a complex nervous system within an embryo. And cultural change is only possible because of the learning capacity of humans. The historical chain reflects a sequence of dependence.

While historical relationships seem straightforward, things are much trickier when we come to relationships of form. Are there fundamental similarities among these various transformations, or does each process operate according to entirely different principles? Attempts to identify similarities in form have been singularly unsuccessful in the past, largely because of two main types of mistake.

One type of error has to do with confounding the familiar with the fundamental. As humans, we are all familiar with the idea of designing and making things, like clothes, furniture, and houses. It therefore seems natural to use the notion of *making* as a general model for how living transformations occur. A tree or frog could arise through making just as a watch is made by a watchmaker. Instead of a human we would need a more powerful divine maker to do the work, but the principle of an external agent being responsible for the design and construction would be the same. This notion of an external agent or divine maker has a long tradition and became embedded in many religions. With the benefit of scientific hindsight we can see where the problem with this explanation lies. The ability of humans to design and create things is a complex feature that arose much later than evolution. Using the notion of making as an overall explanatory principle is employing a complex outcome to account for itself. We only fall into this trap because as humans we are so familiar with making, not realizing how this complex process itself rests on a whole series of other transformations.

It took many generations of scientists to overturn this misconception. A key step was Darwin identifying a simple mechanism (natural selection) that could account for the diverse creatures on our planet. Instead of requiring a divine maker, the evolution of life proceeds inexorably from the way organisms reproduce and interact with their environment. The struggle to establish this viewpoint left a lasting legacy. It established a major divide between our notions about human activities (design and creativity) and the way we think about biological processes such as evolution. These are very different things that we compare at our peril.

We can illustrate another type of error in the search for similarities in form by returning to our game of chess. Take the following conversation between a teacher of chess and a novice pupil:

Teacher: "This piece, the knight, is like a horse because it is able to jump over other pieces."
Pupil: "So, where are its legs?"
Teacher: "It doesn't need legs because the player can make it jump."
Pupil: "How does the player do that?"

Teacher: "By taking the horse and lifting it over the other pieces."

Pupil: "But if the player does the jumping, what is the point of having a horse?"

The misunderstanding arises because the pupil is taking the similarity between a chess piece and a horse too literally. Horses and knights are related only in an abstract or symbolic way. Indeed, it is possible to learn the rules of chess without ever mentioning horses—simply learning the way the knight moves is enough. The game would be exactly the same game if knights looked like teapots instead of horses—all that matters is the nature of the moves they make. Comparisons between knights and horses, if taken too literally, may be a distraction, a misguided analogy that confuses rather than helps.

This type of mistake has repeatedly cropped up in comparisons of different types of living transformation. A case in point is Ernst Haeckel's idea that a fertilized egg reenacts its evolutionary history as it develops. There is no doubt that the way an egg develops is connected with the process of evolution, for it is through evolution that the process of development has arisen. But in trying to take this relationship too far, with development literally repeating the path of evolution, Haeckel ends up missing the mark. He fails to achieve the right level of abstraction and winds up making false and unhelpful comparisons. Edelman's ideas about the brain operating according to Darwinian principles have similarly been criticized for trying to draw too close a comparison among distant processes.

To determine common principles we need to be working at the right level of abstraction. This can only be achieved by having a reasonable understanding of what we are comparing. War and chess can only be compared in a meaningful way after we appreciate what each entails. We don't have to be chess experts or know how to command an army, but we do need an overall idea of how board games and military encounters operate. This allows us to understand both the similarities and the differences between the activities. War and chess are both territorial, but we also understand that there are numerous differences in what a territory means in each case. In war, territory involves land, while in chess it refers to a region of the board.

In the same way, to get at common principles that underlie living transformations we need a broad understanding of how each works. And this has only recently become possible because of advances in our scientific knowledge, particularly in the fields of development and learning. Previous comparisons have been hampered by our not knowing enough about each process, leading us to confound the familiar with the fundamental or to make the wrong types of abstraction. How might we now proceed to address this issue?

Life's Creative Recipe

The sixth-century Chinese art critic Xie He came up with six ingredients that he thought were important for defining the quality of a painting. Roughly translated, these are vitality, brushwork, natural form, color, composition, and copying. All of these aspects, with the exception of color, are illustrated in *Persimmons* (fig. 1), a painting by the thirteenth-century monk Mu Qi. The six ingredients are not completely independent of each other. The vitality of Mu Qi's painting comes partly from the lively brushwork. Similarly, the composition depends on an arrangement of fruits that may have been copied. Rather than providing a set of independent features, Xie He was highlighting some key interacting ingredients that he thought would be helpful in appreciating paintings. Others might come up with a different list of ingredients, so whether or not we like Xie He's choice depends on the extent to which it helps us organize our knowledge

The same applies to the way we organize our ideas. There are many ways of presenting our understanding of processes like evolution, development, learning, and cultural change. In this book I will present a particular viewpoint that emphasizes some overarching features. I will highlight seven key principles that underlie processes from the evolution of bacteria to the workings of our brain. There is nothing mysterious about the number seven here; it is simply that the principles naturally fall into seven categories, just as Xie He found it convenient to define six principles involved in painting.

Our seven principles, or ingredients, and the way they work together define what I call *life's creative recipe*. It is this recipe that lies at the root of how life transforms itself. The evolution of diverse organisms, the development of an egg into an adult, an animal learning of new relationships in

Figure 1. *Persimmons*, Mu Qi (active 1269). Daitokuji, Kyoto.

areas of science treated from a fresh and unifying perspective. Some may feel that areas familiar to them are dealt with in a very selective way. Such selectivity is inevitable in the effort to cover many disciplines and the connections among them, so I hope these readers will forgive my many omissions.

To help convey my viewpoint, I do not keep human creativity at a distance from the scientific story but use it as a recurring theme throughout the book. I often use paintings to illustrate principles, subjects, or ideas I touch upon. The paintings provide visual entry points, and they also serve to remind us of the many perspectives through which we may view things. The artistic and scientific themes eventually come together toward the end of the book. We then see that evolution, development, learning, and culture form a grand cycle, a series of related transformations through which life's creative recipe comes to look back on itself.

ONE

Loops and Lotteries

THE APPLES WE EAT TODAY are not exactly the same as those found in the wild. All cultivated apple trees are thought to have descended from natural populations of *Malus pumila* in the Tian Shan region of Central Asia. These wild apple trees produce fruits that are eaten by large mammals, such as bears, that disperse the seed through their feces. Through many generations of human cultivation and selection, wild apple trees have been transformed to produce fruit more suited to our taste and dinner tables. Darwin proposed that an analogous process of descent with modification was responsible for the origin of all living forms. But while breeding of apple trees depends on artificial selection by humans, Darwin identified a form of selection that occurs without human intervention—the process of natural selection.

Apple trees produce far more apples and seed than ever grow to mature, fruiting trees. So, there is continual competition among seeds in the wild for those that make it through to adulthood. Suppose that among a wild population of apple trees some individuals, say 1 percent of the population, produce apples that are more appealing to bears. Perhaps these apples contain some extra sugar that makes them taste sweeter. Because the bears like them more, the tasty apples would be preferentially eaten and dispersed through the forest. We would therefore expect more of the tasty apples to produce seedlings that grow into apple trees of the next generation. If the tasty trait is inherited, the overall proportion of trees producing these sweet apples increases to say 2 percent. Over several generations, the proportion continues to rise until the population becomes full of trees producing the tasty apples that bears spread more effectively. Through natural selection, the apple trees have become better adapted to reproducing in their environment than they were before.

Scenarios like this provide a general picture of how natural selection can work, but they also raise many questions. The process is often summarized as involving the survival of the fittest. But what do we mean by the fittest? Do we mean those that eventually survive and outcompete, like the trees with tastier apples? If so, then does natural selection boil down to a circular statement—the survival of those that survive? Also, why is it that organisms generally produce far more seed or offspring than can possibly survive? Why not produce about the same number as can survive to adulthood? And where does all the variation in traits, such as apple tastiness, come from in the first place? To answer such questions in a satisfactory way we must look more deeply into how evolution works.

Traditional accounts of evolution by natural selection invoke three main principles. First, individual members of a species vary from one another. Second, individual variation is hereditary to some extent, passed from one generation to the next. Third, organisms multiply at a rate that exceeds the capacity of the environment, with the inevitable result that many die. Natural selection follows as a consequence of these three elements occurring together.

The account of natural selection and evolution I provide here deviates somewhat from this description of only three elements. I instead describe the process through seven principles. Also, I sometimes use terms that do not quite match those traditionally employed. My reason for adopting this distinctive approach is that my overall aim is not only to describe how evolution works, but also to highlight the fundamental ingredients it shares with development, learning, and cultural change.

When we want to understand the essence of a process, it often helps to identify commonalities. Consider the transitions of water mentioned in the previous chapter. If we study how water boils, we might come up with the following notion of what is taking place: as energy is added to the system through heating, the water molecules acquire greater energy and motion until they reach a point where they become more free of each other and expand to form steam. The same explanation applies to ice melting, except that the water does not expand during this transition, but instead contracts. This is because the water molecules in ice are held in an open configuration, which collapses on itself when the molecules are liberated during melting. Ice therefore floats on water—it is less dense than the liquid beneath it. A higher density for the liquid compared to the solid form of a substance is rather unusual—most solids become less dense when they melt. This exceptional behavior of water is informative, however, because it tells us that expansion upon heating is not fundamental to changes in the states of matter. We see what really counts is that molecules gain energy and acquire more freedom of motion; it is these features that are common to different transitions. Identifying commonalities can help us arrive at the heart of what is going on; as distinct from what is more incidental.

Of course there is a danger that this approach could lead us to identify false or superficial similarities. In comparing melting ice with boiling water we might note a change in appearance for both transitions—liquid water looks very different from ice, and steam looks different from liquid water. But while a change in appearance is shared by both transitions, we would hardly call this common feature an explanatory principle. It is more of a description than an explanation, because a changed appearance does not give us any sort of mechanistic insight into what is happening. Similarly, in the determination of unifying principles for living transformations, it is important not to seek commonalities simply for the sake of it. The aim is to look for similar mechanistic elements and interactions, if they exist. And of course I would not be writing this if I did not believe that such elements are there and that they can help us gain a deeper understanding of what transformations in life involve.

To present the commonalities more clearly, I use a set of terms that can be applied to all living transformations. This means that some subjects are placed under slightly unusual headings. For example, in this chapter the subject of hereditary transmission falls under the principle of persistence, and the subject of multiplying organisms falls under the principle of reinforcement. The name for each of these principles has both a general and specific meaning, much as a notion like "composition" has both a broad and particular significance. We may discuss composition in relation to many things, such as paintings, music, or poetry. In all cases, composition refers to an overall arrangement, but what is arranged remains specific to each case—colors on a canvas, musical notes, or a series of words. Similarly, we will see that each of the principles we encounter for living transformations has both a general and specific significance.

In searching for common foundations, I also tease apart some components that are often considered together. This separation clarifies the role of each component, and exposes principles that are sometimes ignored or taken for granted. In this chapter, for example, the notion that organisms reproduce at a rate that exceeds environmental capacity is subdivided into two interacting principles: reinforcement and competition. And in the next chapter we encounter principles such as combinatorial richness and recurrence that are not normally included as explicit ingredients of evolution. You will not find the collection of seven interacting ingredients, which I call life's creative recipe, given elsewhere as an explanation for life's transformations. This is not because I am going to present any new explanatory theories for individual cases. It is because the principles and interactions I describe have been arrived at by taking a perspective that considers all transformations together, rather than each in isolation.

To appreciate the general, we must first understand the specific. Because evolution is the parent of all other living transformations, it is fitting to start with it as our initial example of life's creative recipe. In this chapter I discuss the first four principles of the recipe because they provide the core of natural selection. The remaining three principles will be covered in the following chapter.

Principle of Population Variation

You can never be sure of winning at roulette. If you bet on red, you might expect to win about half of the time, but you cannot predict the outcome of any individual bet. Your chance of winning is actually slightly less than half, because in addition to the eighteen red and eighteen black pockets on the roulette wheel, there is a green pocket, and if the ball lands on green the casino takes the bet. (American roulette tables have two green pockets, and your chances of winning are slightly less.) In the long run, you tend to lose by betting on red; there is only an 18/37 chance of winning, which is just under one-half. It is these small biases that allow casinos to make their money. Casinos are built on statistics.

Statistical reasoning involves two elements. One is the notion of variation. There is a range of possible outcomes for any spin of the roulette wheel and this variation provides the driving force of the game. Variation of this kind is sometimes referred to as being random because of the unpredictability of individual events. But this does not mean that any outcome is possible. Balls do not fall into blue or orange pockets in roulette because there are no such colors on the roulette wheel. Similarly, when tossing a coin we talk about heads or tails, not black or red. Variation always occurs within a particular context.

In addition to variation, the other element of statistical reasoning is the notion of a population. Although you cannot predict what will happen with any individual bet, you know that with many spins of the wheel you are likely to win or lose at a certain rate. When considering many events together (a population), defined features begin to emerge, such as the overall rate of winning or losing. The idea of a population assumes that many events are somehow bound together. The outcomes of spinning a roulette wheel with thirty-eight pockets would not belong to the same population as those from a wheel with thirty-seven pockets. When calculating odds, we also would not mix numbers obtained from rolling dice with numbers obtained from spinning a roulette wheel. Every population is built on some sort of criterion of unity.

Statistical reasoning involves the relationship of two types of units: the individual and the population. These can be thought of as representing different scales or levels. Zooming in, we see individual events that vary (particular spins of the roulette wheel); zooming out, we appreciate the behavior of the population as a whole (the overall rate of losing or winning). With the use of statistics, events at one level can be related to those at another. We can view the behavior of the population as the collective result of variations at the level of the individual. Indeed, statistics provides one of the most powerful tools we have for connecting different levels or scales, for zooming in and out of a problem.

Natural selection is also a statistical process, involving the notions of variation and population. In this case, variation has to do with distinctions among individuals. For the scenario described earlier, some trees produce sweeter apples and others produce less delicious apples. The population is a group of individuals, the collection of interbreeding apple trees in the forest. Natural selection tells us how events at the population level (an increased frequency of trees with tasty apples) may arise through the cumulative effect of variations at the level of the individual (the reproductive success of different apple trees). Let's first look at where this variation comes from.

Heritable variations are carried by DNA, a long molecule made up of two intertwining strands. There are four types of unit, called bases, along each DNA strand, or four letters in the DNA alphabet. These bases are commonly written as A, T, G, and C (for adenine, thymine, guanine, and cytosine). The DNA of a plant or animal, known as its genome, can comprise billions of such bases strung one after another in a sequence. Just as a long piece of text is subdivided into words, so every genome can be subdivided into stretches of DNA, called genes. Humans, for example, have about twenty-five thousand genes in their genome. Each gene contributes to particular characteristics of an individual, such as the taste of its apples or the color of its eyes.

Slight changes can sometimes occur in the DNA sequence, such as C being replaced by a T at a certain position. Mutations of this kind may be triggered by radiation hitting the DNA, or by an error that takes place when the DNA molecule is copied. If a DNA mutation occurs in a cell giving rise to eggs or sperm, it may be passed on to the next generation, thereby introducing a mutant gene into the population. And because the features of an individual depend on its DNA sequence, this may lead to heritable variations in the population, like variation in the sweetness of an apple or in eye color.

There are, however, limits to the type of variation we expect. There are no mutations that result in apples with blue eyes. This is because variation always arises in a particular context, in relation to a particular DNA sequence or genome. For an apple tree genome, mutations may have the effect of producing different types of apples, but not of resulting in human features like eyes. The variation is defined by the genome and organism we begin with, by the context.

Whether a mutation occurs at a specific location in the DNA at any given time is unpredictable. Nevertheless, in a population of many DNA molecules we can talk about the probability of a mutation occurring at a particular location. This probability is normally very low, around one in a billion cell divisions. Even so, because DNA is so long and there are many individuals in a population, mutations are always cropping up. Although each mutation is a rare event, they continually arise in populations, providing genetic variation.

In addition to the process of mutation, genetic variation is also enhanced by sexual reproduction. The genome is normally not just one long piece of DNA, but is broken up into several separate lengths, called chromosomes. For sexually reproducing organisms, these chromosomes are in pairs, one donated by each parent during the act of reproduction. You, for example, have twenty-three pairs of chromosomes in your genome. Twenty-three of these chromosomes were inherited from your father and the other twenty-three from your mother. The pairs can get shuffled, like a pack of cards, from one generation to the next when sperm or eggs are produced. As a result of this process, chromosomes come together in new combinations when an egg is fertilized, increasing genetic variation in the population. New genetic combinations may also arise through exchanges between chromosomes. Chromosomes may break and be rejoined during production of sperm or eggs, such that a section of one chromosome is swapped with the corresponding section of its partner. This process of swapping (recombination) further increases genetic variation in the population.

Heritable variation is therefore continually generated and is a basic feature of all organisms. This variation provides the first element used in statistical reasoning. We now come to the second element—population behavior.

Natural selection depends on variation in a population comprising many individuals. Consider the example of the apple trees, in which the proportion producing sweet apples gradually increased because their seeds were dispersed more effectively by bears. It is important here to consider what happens to the population as a whole rather than to any particular individual. For example, a particular tree with sour apples may leave more offspring than another tree that bears sweet apples. Perhaps this sour-apple tree happens to grow in a particularly fertile patch of the forest, or maybe it is located next to where a family of bears lives. Whatever the explanation, the success of this individual tree is not central to natural selection because it is reproductive behavior over the entire population that counts. Across the population as whole, trees with sweet apples are on average more successful in reproducing than trees with sour apples, leading to an increase in the proportion of trees with tasty apples. Similarly, a particular rate of winning or losing at roulette emerges from a group (population) of bets, not from any individual bet. Natural selection is about rates of survival and reproduction in a population rather the fate of one or two individuals. It is a statistical process based on population variation.

Principle of Persistence

On Thursday, September 12, 1940, four French boys from the town of Montignac were exploring a hole left by an uprooted tree in the countryside. They had brought along a lamp because they had heard a rumour that the hole led to an underground passage. After plunging into the hole and lighting their lamp, the boys discovered a series of subterranean caves with the walls covered by paintings of animals. They had stumbled into one of the most stunning displays of prehistoric art in the caves of Lascaux. These paintings have survived in remarkably good condition for more than ten thousand years (fig. 2).

The long-term survival of these paintings reflects the stability of rock and pigment. It is a testament to the strength of molecular forces that keep the rock together and the paint adhering to its surface. However, even if there was a calamity and the caves of Lascaux were destroyed, we would still have an idea of what the paintings looked like. This is because of another form of persistence. The photographic negatives, positives, and electronic images of Lascaux provide another means of preservation. This form of persistence also depends on molecular forces: those that prevent the molecules in the photographs or electronic devices from flying apart. But there is the added benefit of safety in numbers here, because the information can be extensively copied and propagated. For this reason, millions of people who have never visited the caves can nevertheless appreciate the drawings through the many photographs and images that are available.

These two forms of persistence (molecular cohesion and copying) are also central to Darwinian evolution. If organisms were not held together by molecular forces but were continually flying apart, evolution could not take place because there would be no organisms to speak of. There would also be no DNA. The bases in the DNA are held together in a se-

Figure 2. Cave painting from Lascaux, France, 15,000–10,000 BC.

quence by forces that maintain the backbone of the molecular chain. Were it not for these forces, there could be no DNA sequence, and no genes.

The DNA sequence can also persist through replication (copying). DNA of each chromosome is made up of two strands that match each other, in the same way that photographic negatives and positives are complementary. Whenever there is an A on one DNA strand, there is a matching T on the opposite strand, forming what is called an A-T base pair. Similarly G and C are always opposite each other, forming a G-C base pair. In this way, the base sequence of one strand contains all the information needed to specify the sequence in the complementary strand—the sequence AGCT on one strand specifies TCGA on the other, just as a photographic negative is always matched to a photographic positive. Replication takes place when the two DNA strands are pulled apart and each acts as a template to specify its complement, allowing two DNA molecules to be made from one—the positive strand specifies a new negative partner, while the negative strand specifies a new positive partner. This process of copying happens every time a cell divides, so that both daughter cells end up with copies of the genome. This DNA copying also allows information to be propagated and passed on from one generation to the next.

As we have already learned from the principle of population variation, the DNA sequence is not completely immutable. Alterations in the sequence may occur through damage or through errors in the copying process. If such a mutation renders an organism less able to survive and reproduce, the mutant version will most likely be eliminated from the population by natural selection. On the other hand, a mutation may occasionally lead to improved reproductive success; in that case, natural selection would favor the spread of the mutant version in the population. The ability of variations to persist through replication makes all of this possible.

There has to be a balance between variation and persistence for evolution to take place. Variation provides the raw material for natural selection, while persistence allows information from one generation to be passed on to the next. Variation without persistence would mean that changes could not be maintained and built upon during evolution. And persistence without variation would bring evolution to a standstill.

The balance between persistence and variation is not only a prerequisite for evolution; it is something that itself has evolved. Replication of DNA, for example, depends on proteins which separate and copy the DNA strands. If these proteins do a sloppy job and make lots of errors when copying the DNA, many mutations would be introduced. This would produce a shift toward variation over persistence. However, too many errors would lead to the DNA being copied ineffectively. Its sequence would begin to degenerate, just as a bad photocopying machine leads to the loss of information from an image. This negative impact of sloppy copying means that the proteins responsible would tend to be disfavored by natural selection, shifting the balance back in favor of persistence over variation. Over billions of years of evolution there has been continual interplay between persistence, variation, and selection. The balance between variation and persistence we see in populations around us today is a product of this long evolutionary history.

The interplay between persistence, variation, and selection provides the first of several feedback loops we encounter in this chapter. It illustrates how the various principles I describe are not completely independent of each other. Variation and persistence are not only ingredients of evolution, they are themselves modified through the evolutionary process.

Principle of Reinforcement

Population variation and hereditary persistence provide the backdrop for evolution. But on their own they do not provide a driving force for evolutionary change. Such a force emerges when we consider a particular class of variations, those that influence reproductive ability.

At the heart of this process is the law of compound interest. Suppose you have a population of one hundred balls. The balls have the ability to duplicate occasionally, so that two new balls can sometimes arise from a single ball. Imagine that over a period of one hour these duplications lead to an average increase in the number of balls of 10 percent. This means that after one hour we have around 110 balls. There are now slightly more balls that can potentially duplicate. Consequently, there will be a slightly greater increase in the number of balls in the next hour—eleven balls on average (10 percent of 110), resulting in a total of 121 balls. The number of balls continues to increase in this way (according to the rules of compound interest), and after ten hours we have about 260 balls. After one hundred hours we have more than a million balls.

Now suppose that the balls are two different colors, black and white. We can assume the color of the ball is transmitted faithfully when a ball duplicates, so that a black ball gives birth to black, and a white ball to white. If we start with an equal number of black and white balls and they both replicate at the same average rate, then their proportion remains roughly equal as the population increases in size. The situation is different, however, if the colors reproduce at different rates. Imagine that the black balls increase at an average rate of 11 percent per hour compared to a rate of 10 percent for the white balls. Starting with a population of one hundred black and one hundred white balls, on average we expect to end up with 111 black and 110 white after one hour; then the black balls comprise slightly more than 50 percent of the population (about 50.2 percent). After ten hours there are about 280 black and 260 white balls, so the proportion of black has increased to around 52 percent. After one hundred hours the population has grown to millions, and the proportion of black balls is about 70 percent. The law of compound interest ensures that the slight difference in reproductive rate between the black and white balls leads to a numerical difference that is reinforced hour by hour, so that black balls outnumber white balls by more and more.

Reinforcement lies at the heart of natural selection. Genes, or more strictly speaking, gene versions, that help an organism reproduce will tend to be better represented in the next generation. Being even slightly more abundant, these gene versions start with an advantage that can multiply in the following generation. Repeating this process again and again, a gene that favors reproduction can continually reinforce its numbers, and increase in the population.

Reinforcement alone, however, cannot result in one gene version replacing another. Even if we let our population of black and white balls continue growing for many days, so that more than 99 percent are black, there would still be a vast number of white balls present. This is because no balls are lost; we are only gaining more and more. Reinforcement can lead to an increase in proportion but not to the replacement of one gene version by another. To see how replacement may occur, I need to introduce a further principle: competition.

Principle of Competition

The idea of natural selection first occurred to Darwin shortly after he returned from his five year odyssey on the *HMS Beagle*, the ship on which he served as naturalist. As a result of his trip, Darwin became convinced that species were not fixed but could be modified over time. But he did not know of a mechanism that could explain how species change and adapt. Then, in October of 1838, two years after his return, he was reading a book on population growth by Thomas Malthus. Malthus had pointed out that if the human population continued rising according to the laws of compound interest, then the population size would eventually outstrip the food supply, and struggle and starvation would inevitably follow. This idea struck home:

> I happened to read for amusement Malthus on *Populations*, and being well prepared to appreciate the struggle for existence which everywhere goes on from long-continued observation of the habits of animals and plants, it at once struck me that under these circumstances favourable variations would tend to be preserved and unfavourable ones destroyed. The result of this would be the formation of new species. Here, then, I had at last got a theory by which to work.

Darwin realized that if all the offspring of plants or animals survived, population numbers would soon pass beyond the limits of what the environment could sustain. The result would be a continual struggle or competition for existence, with many individuals dying. Those better adapted would have an improved chance of survival, resulting in a natural form of selection. Reading Malthus was important for Darwin because it brought together two ideas—populations tend to grow according to the law of compound interest, and this growth is eventually limited because the environment's resources are finite. It was no accident that Alfred Russel Wallace was also contemplating Malthus when he came up with idea of natural selection about twenty years after Darwin (but before Darwin had published). In both cases, it was the confluence of ideas about reproductive increase and environmental limitations that provided the seeds for the formulation of natural selection.

The notion of competition is often linked with that of reinforcement. This is because, as Darwin and Wallace realized, competition often arises through the pressure of large numbers. But to get a clearer understanding of natural selection, it first helps to separate these two notions and consider what competition represents on its own, without reinforcement. Considering competition alone will also help us understand why the theory of natural selection is not a circular argument involving the survival of those that survive. To appreciate the effects of competition without reinforcement, let's return to our example of black and white balls.

Imagine a lottery machine that contains an equal number of black and white balls that are continually being mixed (fig. 3, *left*). Say there are five

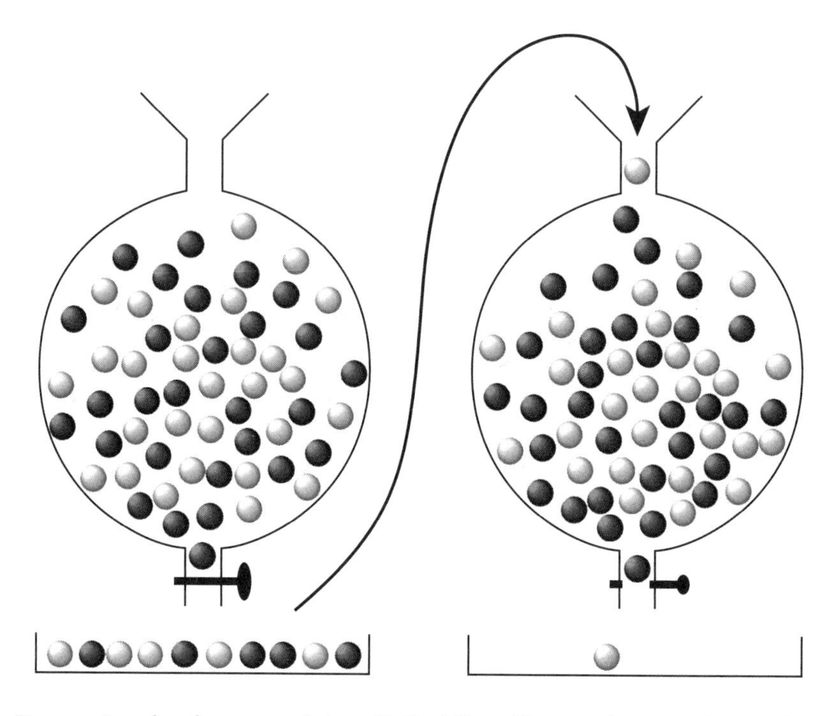

Figure 3. Sampling from a population of balls, followed by reproduction and resampling.

hundred balls of each color, so our population comprises one thousand balls altogether. A few balls are sampled every so often from an outlet at the bottom of the machine. After a while, the outlet is closed and we obtain a limited number of balls in our sample, say one hundred. We call this the sample population. Because the sample is smaller than the population as a whole, we can say that the balls in the lottery machine are competing with each other for a place in the sample. By using the word "competition" here, I don't mean that the balls are actively struggling with each other, but only that there are a large number of candidates for a limited number of places.

Now suppose the one hundred balls in our sample population replicate and make lots of copies of themselves, so that we eventually end up with a thousand balls again. These balls are fed into an empty but identical lottery machine to give a new population of one thousand balls. We are more or less back to where we started, having gone through a cycle of sampling and replication. We can now go through a whole other cycle, or generation, taking a sample population of one hundred balls from the new machine, and then replicating these balls to provide a thousand new ones that are put into yet another empty lottery machine. The process of mixing and sampling can be repeated again and again, over more and more generations. What happens to such a system over time?

Let's first consider a situation where there is no difference in the properties of the black and white balls. That is, they compete equally well for a place in the sample (much like a human competition with many candidates of equal ability competing for a limited number of places). If we start with black and white balls in equal numbers, you might expect that the proportion would remain roughly the same over many generations. Surprisingly, however, after about one hundred generations, the population will most likely contain only black or only white balls. This outcome has to do with the way that chance variations can build up. In our first sample of balls, for example, we might not find exactly fifty white and fifty black balls, but we might instead find slightly more of one color. Suppose we find fifty-three black balls and forty-seven white balls. Following replication, we have about 530 black balls and 470 white balls in the next lottery machine. The fluctuation in our sample has been incorporated into the lottery population. This means that the probability of getting black balls in our next sample is now slightly more than half (53 percent). Fluctuations in our next sampling event may further enhance or perhaps reduce this difference. In this way, the proportion of black or white balls can drift up or down, with fluctuations repeatedly passed from one lottery machine to the next. After about one hundred generations of fluctuations accumulating like this, the population may eventually drift to only one color. When that happens, there is no going back and the population remains a single color, say black. The population has then become what we call *fixed* for black balls. Just from the effects of repeated sampling (repeating competition for a limited number of places), the population has evolved from including balls that are half black and half white, to becoming fixed for one color. This happens even though any particular ball, black or white, has exactly the same chance of getting into the sample each generation (every participant in our competition has exactly the same ability and thus the same chance of getting through).

Compare this outcome to what takes place in a single sampling event. If you were to sample one hundred balls from an equal mixture of black and white balls, the probability that all would be one color is about 1 in 10^{30}, or one in a thousand billion billion billion. If you were only to take single samples, you would have to keep trying for longer than the age of the universe to stand a chance of getting a single color of balls. Yet with our previous scheme of repeated rounds of sampling and replication, ending up with one color is almost inevitable and takes only about one hundred generations. (The number of generations necessary is proportional to the size of the sample population.) By introducing a limited sample (competition for a limited number of places), and then allowing each generation to build on the previous one, we have changed the character of the process. Now there is a natural tendency for one form to take over. There is an equal chance that the dominating color will be black or white, but one will certainly replace the other eventually.

Life is also a lottery. When wild apple trees in the forests of the Tian Shan produce their vast number of apples, only a small proportion of the seed in these fruits grow into mature trees that produce seeds for the next generation. This is because the Tian Shan region is of finite size and can support only so many apple trees. The mature trees are equivalent to our sample population from the lottery machine. If we imagine that each tree produces one million seeds each generation, then on average, only one in a million seeds will grow into a tree. In this case, our sample population would be one millionth of the population in the lottery machine. There is intense competition for relatively few places. And like the lottery cycle already described, the process repeats itself generation after generation.

The black and white balls are equivalent to different gene versions in the apple tree population. Genetic variation of this kind is continually being introduced into the population through mutation. As new versions crop up, some may drift to higher levels in the population because of the repeated sampling (competition) that occurs in each generation. Such genetic drift due to limits in population size takes place even if the gene versions have no effect on seed survival and reproduction. Eventually some variations become fixed in the population, just as we saw for repeated sampling of the balls.

The process of genetic drift shows how the introduction of a population limit can have a dramatic effect on what takes place over multiple generations. In the previous section on the principle of reinforcement, we saw how the law of compound interest can lead to a gradual increase in the proportion of a gene version if it has a reproductive advantage over another. However, as the other gene version was not eliminated, one version never took over completely. By introducing a limit on our reproducing population, some gene versions can rise to complete supremacy. This happens even if the competing gene versions have exactly the same reproductive rate. Competition itself, irrespective of the merits of each version, has allowed our population to evolve and change over time.

We are now in a position to see why the theory of natural selection does not involve circular reasoning. As we have seen, one form can survive and replace another in a population just through the effects of repeated sampling (genetic drift). This is important because it shows that a population can evolve, changing its genetic makeup, in the absence of natural selection. Evolution and natural selection are not one and the same. When discussing natural selection, *fittest* does not designate those that eventually survive in the population. If it did, then we would say that a black ball was fitter than a white ball, even if it increased to ascendancy through repeated sampling, without any inherent advantage over white. Fitness, however, has an altogether different meaning. It does not denote those that eventually survive but involves the introduction of a bias into the competition, as we will now see.

Combining Principles

So far, we have assumed that black and white balls in our lottery machine have the same ability to reproduce and compete for their place in the sample population. They therefore have an equal chance of eventually taking over the population. Let's now look at what happens if we introduce a difference in their properties. Suppose, for example, that black balls replicate at a higher rate than white. This is the same situation as described for the principle of reinforcement, but we are now combining it with the idea of a lottery machine with limited places in the sample (competition). After the sample has been taken, the greater reproductive ability of black balls means that their numbers will tend to increase more than white. They are therefore likely to contribute a greater proportion to the next generation, creating a bias in favor of black. This bias is reinforced each generation; over several generations of sampling it is more likely that the population will become fixed for black rather than for white.

By combining reinforcement (a greater reproductive rate for black balls) with competition, two things have happened. First, it is now more likely that the population becomes fixed for black balls rather than for white balls: the system has become biased toward the color that reproduces better. Second, the rise in proportion of black balls happens more quickly than it would if the two colors had the same reproductive ability. This is because the proportion of black balls is given a boost each generation, increasing the rate at which the percentage of black rises. The rate of increase in proportion of black balls eventually drops as they become more prevalent in the lottery machine. This is because as the proportion of white balls decreases, the black balls start to compete more and more with other black balls for places in the sample. The success of the black balls ultimately causes a reduction in black's rate of increase. Nevertheless, it takes less time overall for black balls to become fixed than when they have no advantage over white balls. These two consequences, a biased outcome and a shorter time for fixation, are often used by evolutionary biologists to infer that natural selection rather than just genetic drift is taking place in a population.

There is another way in which the black balls' chances can be favored. Suppose that they are slightly heavier than the white balls and tend to stay closer to the bottom of the lottery machine, resulting in a greater chance of falling into the sample. The black balls are effectively outcompeting the white balls for their place in the sample population. (This is similar to a human competition where the candidates are not equally able, but some are better than others and are more likely to win.) More black balls in the sample means that they contribute more to the next generation, even if black balls replicate at the same rate as white. The advantage of black is reinforced in each subsequent generation so it is now most likely that we will end up with all black balls. Changing the ability to get a place in the sample has a similar effect to changing the reproductive rate. It is another form of reinforcement, but one that arises through the process of competition itself.

The same principles apply to our population of apple trees. If one gene version promotes the ability of seeds to survive and grow into a mature tree, then that gene version tends to increase in the population; it has a selective advantage, or higher fitness, equivalent to a heavier ball in our lottery machine. Another type of selective advantage occurs when the gene version improves the apple tree's ability to reproduce. A version that makes the apple blossom more attractive to bees, for example, might lead to production of more seed. This reproductive advantage would also be reinforced every generation, increasing the gene version's chance of becoming fixed.

These two forms of reinforcement, an increased ability to survive and an increased ability to reproduce, distinguish selection from genetic drift. With genetic drift, gene versions have equal chances of becoming fixed. With selection, the versions have different chances of becoming fixed; those that promote survival and reproduction are favored. This is how natural selection leads to a match between organism and environment; it favors features that increase survival and reproduction.

The interaction between reinforcement and competition lies at the heart of natural selection, and is summarized in the diagram in figure 4. The positive feedback loop represents reinforcement; if a gene version promotes reproductive success, it tends to boost itself to higher levels with each generation. The negative feedback loop represents competition and arises through environmental limitations that constrain population size. A population limit has the consequence that the rise in frequency of one gene version is at the expense of another; this may eventually lead to elimination of the competing gene version. A further consequence of a population limit is that a gene version competes against itself as it becomes more prevalent in the population. The gene version becomes a victim of its own success; as it proliferates in the population, it is more likely that competing individuals have the same gene version, so its rate of increase diminishes. By the time the gene version takes over completely, all the individuals will be equally able to compete because they share the same advantage. The overall result is that all individuals are better able to survive and reproduce. An adaptation has become established in the population.

The double feedback loop of reinforcement and competition helps explain why organisms normally produce more offspring than can survive. If we start with a population that produces the same number of offspring as can be sustained by the environment, there would be no competition because all could survive. But suppose a gene version arises through mu-

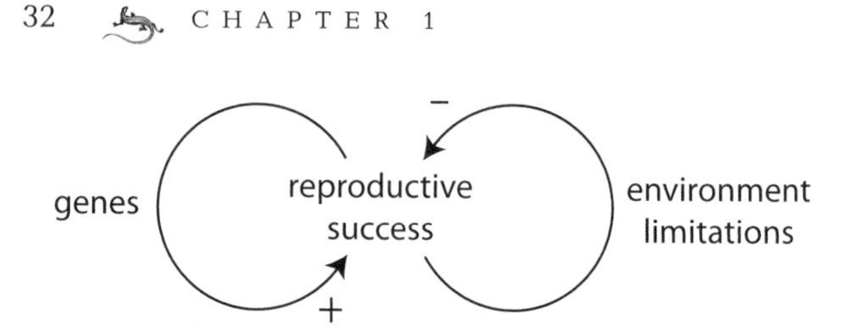

Figure 4. Interplay between reinforcement (positive loop) and competition (negative loop) for natural selection.

tation that increases the reproductive rate. The gene version will likely spread in the population through reinforcement (positive loop in figure 4). As this happens, the population reproduces more effectively, leaving more offspring than before. Then the population numbers will exceed the environment's limits, resulting in competition for limited resources (negative loop in figure 4). The increase in reproductive success ushers in the competition that limits it. Our double feedback loop is not only able to drive evolutionary change, but it also sustains itself.

This double feedback loop involving reinforcement and competition is the engine that lies at the core of natural selection. The engine is fueled by a balance of population variation and persistence. Without variation there would be nothing to keep the engine running, and without persistence it would fail to drive change. We have also seen that there is a further feedback loop from the engine to its fuel: the accuracy of DNA copying, and thus the extent of variation, is itself modulated by natural selection (pages 22,23).

Although these feedback loops are implicit in many accounts of evolution, we have been able to view them more clearly by first separating out their key ingredients. We can then appreciate how natural selection is driven through an interplay between four principles—population variation, persistence, reinforcement, and competition. In later chapters we will see how similar principles and feedback loops lie at the heart of other types of living transformations. They provide a common form that crops up under many different guises, similar to the way that chess and war have a common form even though they differ in many respects. But for now I want to continue with the story of evolution by looking at some features we have skipped over or taken for granted.

Throughout this chapter, we have been assuming that natural selection operates within a particular context, a population of individuals. We can discuss natural selection acting within the apple tree population, favoring one type of apple over another; or acting within the population of bears, favoring some bears over others. But we don't talk of natural selection favoring an apple tree over a bear. Even if there are more apple trees than bears, this does not mean that natural selection is favoring one over the other. Natural selection instead concerns the relative success of individuals within the same population rather than absolute numbers. This begs the question of why individuals exist to begin with, and why they are organized in so many diverse populations. Why are some populations made up of lots of individual apple trees, and other populations made up of bears? To answer these questions we will look at three further principles, the remaining components of life's creative recipe.

TWO

From Genes to Ecosystems

According to the art dealer Ambroise Vollard, having a portrait painted by Cézanne was a long and drawn-out experience. As the painting reached completion, Vollard made the mistake of commenting on two areas where it seemed unfinished:

> For someone who has not seen him paint, it is difficult to imagine the extent to which, some days, his progress was slow and painful. In my portrait, there are two little spots on the hand where the canvas was not covered. I pointed them out to Cézanne. He replied: "...I may be able to find the right color with which to fill in those blank patches. But understand a little, Monsieur Vollard, if I put something there at random, I would be forced to begin my painting again, starting from that point."

Cézanne never changed those two patches—they can still be seen as white flecks at the base of the middle finger in figure 5 (plate 1)—so Vollard was saved from a further series of sittings. Cézanne was keenly aware of the interplay among colors on a canvas. Each patch would have to be painted in precise relation to the others, as it was through the interaction of the colors that the harmony of the picture as a whole emerged. A nonharmonious change, such as placing the wrong color on the hand, might spell disaster.

Like paintings, organisms are spatial units in which the various parts interact to form an integrated whole. Our ability to breath depends on the interplay between our nervous system, muscles, skeleton, and lungs.

Figure 5. *Portrait of Ambroise Vollard*, Paul Cézanne, 1899. See plate 1.

The function of our lungs depends on the composition of the mucus that lines its walls. The composition of the mucus depends on the proteins that transport negatively charged chloride ions. Changes in just one element of this integrated system can have disastrous consequences. Patients with cystic fibrosis have difficulty breathing because they carry a mutation in the gene needed for chloride transport. It only takes one change out of the three billion base pairs in our genome to cause the disease. The functioning of every individual depends on the integration of many different components.

In the previous chapter we took it for granted that the living world is organized into spatial units called individuals. Variation among these individuals from a population allows natural selection to work. But why is life organized in this way?

Principle of Cooperation

You might think that because of the principle of competition, natural selection would lead to ruthless selfishness. But when faced with severe competition, it is often advantageous to combine forces and work together, as long as this is of mutual benefit. This type of interaction emerges chiefly when components are connected or share the same space.

Consider a unicellular organism. Its DNA comprises long sequences of base pairs strung together, forming chromosomes. The DNA can get recombined from one generation to the next by random breaking and rejoining of chromosomes. During this process, pieces of DNA are swapped between chromosomes (recombination). Because the breaks occur at only a few places along the DNA molecule during sexual reproduction, bases that are close to each other are unlikely to undergo recombination. This is for the same reason that if you are standing in a long queue, you are unlikely to be separated from your nearest neighbors if the queue is randomly split in two. So, in a DNA molecule, bases close to one another tend to stay together from one generation to the next. Consequently, if a mutated base at one position in the genome promotes reproductive success, and thereby its own spread in a population, it will likely aid the spread of any bases that were near it, by carrying them along. Similarly, a base stands to benefit if any of its neighbors happen to increase reproductive success. There is a mutual incentive for nearby bases to cooperate, because their close physical connection means that they are more likely to be inherited together. By incentive to cooperate, I don't mean that one base is thinking about the other, but that what is beneficial to one, in terms of reproductive success, is also likely to be beneficial to the other. In such a situation, natural selection favors outcomes in which nearby bases effectively work together to ensure reproductive success. Of course nearby base sequences could also interact negatively to reduce reproductive success, but in that case they are less likely to become fixed through natural selection, and therefore do not become prevalent. "Cooperation" is used to refer to collaborative outcomes without any psychological intent, just as "competition" was used in the previous chapter to refer to a situation where there are many entities, only a few of which attain a limited number of places.

One form of cooperation among bases is to code for a protein. Proteins are large molecules, many of which act by increasing the rate of particular chemical reactions, such as the breakdown of glucose or the fixation of carbon. Each protein is composed of a string of molecular subunits called amino acids. There can be twenty different types of amino acids in a protein, or twenty letters in the protein alphabet. The properties of a protein depend upon the precise order in which its amino acids are strung together. This amino acid sequence in turn is determined by a stretch of DNA sequence (a gene) in the genome. Every type of protein has a corresponding gene that codes for it. The genome codes for many thousands of proteins, so each of these DNA sequences represents only a tiny fraction of the total DNA.

A protein is a cooperative or joint effort because it is coded for by a string of nearby bases in the DNA, not just a single base. One base alone cannot code for a protein; a sequence of bases in a gene is required. There is another relationship between the gene and protein—both reside in the same cell. The outer membrane of the cell prevents the protein from drifting too far away from the gene and ensures that the fates of gene and protein are intertwined. This means that if the protein promotes the cell's survival and reproduction, it also increases the reproductive success of the gene that coded for it; if the protein was completely free to move away, it would be of no direct benefit to the gene. We have two intimate physical relationships—bases within the same gene are linked together, and proteins are also kept nearby due to containment by the cell membrane. This spatial intimacy locks the fate of these different components together; what favors one in terms of reproductive success also favors the other. Of course this also means that if one component reduces reproductive success it will bring the other down with it. But such changes will not be favored by natural selection (they will be selected against). Variations that increase reproductive ability are the ones that tend to get established in populations. Being in the same boat encourages mutual assistance and sharing of benefits.

When we consider the many other genes from our unicellular organism, the same also applies. All of these genes reside together within the same cell. So it pays for them to cooperate by their proteins working together. The intimacy between different genes, however, is not quite as strong as between bases from the same gene. This is because DNA sequences that are further apart from each other in the genome are more likely to be shuffled and recombined as they are passed on from one generation to the next. It is therefore likely that a gene version that arose in one individual will break away from its original neighbors and join another gene version that arose in another individual. Each gene version is assessed by natural selection in relation to the range of genes it encounters in the population, rather than only in relation to its original neighbors. Cooperation is still paramount because each gene needs to work effectively with the other genes in a cell in order to promote its reproductive success, but the drive for cooperation is less intense than for nearby bases.

In multicellular organisms, cooperation is extended to all cells in an individual. As a fertilized egg grows and divides, the DNA is faithfully copied so that, with a few exceptions, all the cells in the adult contain a copy of the same DNA sequence. Such an arrangement promotes collaboration—if cells within the same individual cooperate, the spread of their shared genes will be favored in the population through natural selection. Again, physical proximity is of central importance. If cells moved off in different directions after fertilization, they would no longer be able to assist each other effectively, or share any benefits that might accrue.

We tend to take the notion of individuals for granted. It seems obvious that organisms exist as spatially organized packets. However, this individuality is not a given, but is the outcome of two underlying factors. The first is that nearby entities usually interact more intensely than those at a distance from one another. This is a fundamental property of matter, reflecting the decrease in strength of physical forces, such as electrical attraction or repulsion, with distance. The second is that in a competitive setting, it can pay to cooperate if fates are intertwined. In the context of natural selection, cooperation is promoted if entities are in the same boat, so that helping each other is made easy and reproductive benefits can be shared. This leads to collaborations among components that are closely connected in space, be they DNA bases, genes, proteins, or cells. Proximity helps bind activities and fates together. The first living organisms may have comprised relatively crude collaborations, but over billions of years of evolution these have been modified through natural selection, resulting in the individuals we see today.

Because life forms are organized as cooperative units, the significance of a change in one component can only be judged by considering how it interacts with the others. The effect of changing one base depends on its surrounding DNA sequence. The base on its own conveys very little, just as a single letter from a word carries little meaning in isolation. The same applies to amino acids—the consequence of changing an amino acid depends on how it interacts with all the other amino acids in the protein. Similarly, the effect of changing the protein depends on how it interacts with the constellation of other proteins in the individual. The notion of natural selection acting on a base or gene in isolation makes little sense because bases and genes only exert their effects by interacting with multiple components. These interactions may take place at many levels: adjacent bases in a gene, different components within a cell, groups of cells in a tissue, or various organs of an individual. This is similar to the colors that interact at many levels in Cézanne's painting.

We have seen how cooperation emerges as an outcome of natural selection. But cooperation can also be viewed as an ingredient of natural selection. For without cooperation there would be no individuals or genes for natural selection to act on. Competition and cooperation feed off each other. Competition leads to cooperative spatial units and these in turn provide the assemblies that drive further competition. This continual feedback between competition and cooperation is a fundamental feature of evolution by natural selection, yet it is often overlooked. Natural selection is typically described as being primarily about competition rather than cooperation. This is the path that I also followed when introducing natural selection in the previous chapter. Yet such descriptions take the existence of cooperative units, like individuals or genes, for granted.

Grounding natural selection purely in competition ignores the feedback loop that links competition with cooperation. By their nature, feedback loops do not have a clear starting point; giving competition primacy makes an artificial break in what is actually a cycle of interdependence. Cooperation is a principle of evolutionary change that is just as fundamental as competition.

Principle of Combinatorial Richness

In one of his portraits of neurology patients, Oliver Sacks describes the case of Mr. I., a painter who became color-blind at the age of sixty-five following a car accident. Prior to this he had been a successful artist with a keen appreciation of color. After the accident, tomato juice looked black and his brown dog appeared gray. "As the months went by, he particularly missed the brilliant colors of spring—he had always loved flowers, but now he could only distinguish them by shape or smell. The blue jays were brilliant no longer; their blue, curiously, now seen as a pale grey." Mr. I. felt an unbearable sense of loss and suffered an almost suicidal depression during the first weeks following his accident.

We take the richness of the world for granted and only fully appreciate its diversity when a particular feature is removed. Of course color is just one attribute of many that we take in. Without color vision Mr. I. could still observe the coming of spring, the shape and smell of flowers, and the flight of a bird. In doing so, he was taking in the world's complexity at a series of different scales, from the earth's orbit around the sun, to the shape or movement of nearby objects, to the chemicals emitted by a flower.

But suppose the earth was just a boring, monotonous lump of matter without any variation to speak of. Regardless of what senses we had, the world would seem dull because there would be nothing to engage us. The consequences, however, would go much further; we would not even be here to contemplate such a tedious world. This is because the complexity of living organisms (including ourselves) is related to the complexity of the world they live in. Natural selection can only lead to adaptations if there are features in the surroundings to adapt to. Our ability to contemplate what is around us has only evolved because there is an environment worth contemplating. If the world was truly dull, then we would be dull too.

Of course, our world is not so boring. The earth contains many different atomic elements that can combine with each other according to the laws of physics and chemistry. Although there are only about one hundred different types of atoms, the number of ways they can combine together is far more numerous. Hydrogen and oxygen can combine to form molecules of water (H_2O). Sodium and chlorine can associate to form salt. Moreover, these molecules can themselves combine with each other in various ways. Water molecules can come together to form ice, snow, rivers, vapor, or clouds. And salt can combine with water to result in a solution of sodium chloride. A vast range of arrangements can be produced simply by joining a few elements in different ways. I call this the principle of combinatorial richness. It is this principle that makes our world such a rich and interesting place.

There are consequently plenty of challenges for organisms to grapple with in the world around them. Organisms meet these challenges through their internal wealth of molecular arrangements. This wealth of possibilities again arises through the principle of combinatorial richness. To see how, let's consider a rather unrealistic situation in which we could vary the length of an organism's DNA at will, while still allowing the organism to reproduce. Suppose we whittle the DNA down to its smallest length, only one base pair long. There would then be only two possible types of genome, an A on one DNA strand opposite a T on the other, or G on one strand opposite C on the other. No matter how interesting the world might be, such organisms would only have the choice of two possible DNA configurations.

If we allow genomes to be two base pairs long, there would be more possibilities. For example, we could have organisms with AA, AG, or AC along one strand of their genomes. We can represent all the possibilities with a diagram in which the base in the first position varies along the horizontal axis and the base at the second position varies along the vertical axis (fig. 6). The resulting square indicates sixteen possible organisms (4×4, or 4^2). However, because each of these base sequences represents only one of the two DNA strands, some of them represent the same genome sequence (AA, for example, is the same as TT viewed from the

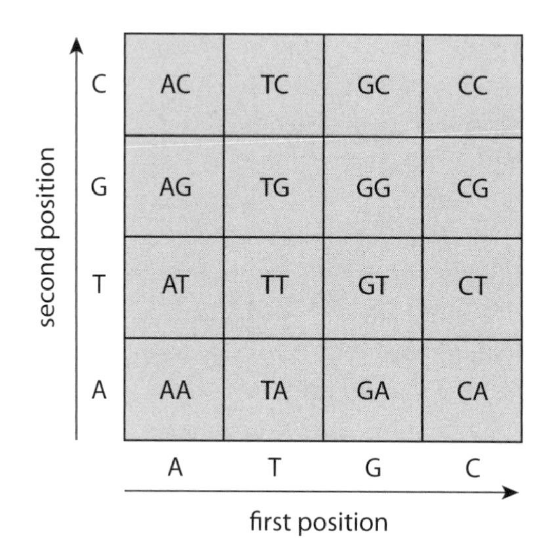

Figure 6. Possible sequences for one strand of a DNA molecule two base pairs long.

other strand). This means we have to divide the total number of combinations by two, giving eight ($4^2/2 = 8$) different genomes. This compares with the two possibilities for genomes that are a single base long. By simply adding the same type of unit, another base, we have multiplied the possibilities by four.

A genome three base pairs long, with sequences like AGC and TAG on one strand, would yield a total of thirty-two possibilities: ($4 \times 4 \times 4$) divided by two ($4^3/2$). By adding another base, the number of possibilities is again multiplied by four. The various possibilities can be represented in three dimensions by a cube; each axis represents the identity of the base in the first, second, or third position (fig. 7). The principle of combinatorial richness is becoming evident; as more bases are added, the number of possible combinations increases very quickly.

In genomes that are four base pairs long, we would find 128 possibilities ($4^4/2$). We would then need four axes or dimensions to illustrate the possible combinations. Unfortunately, this scenario of four-dimensional space takes us beyond the limits of what we can easily imagine. Nevertheless, we could think of this space as a glorified cube with an extra dimension thrown in. Each position in this four-dimensional hypercube would

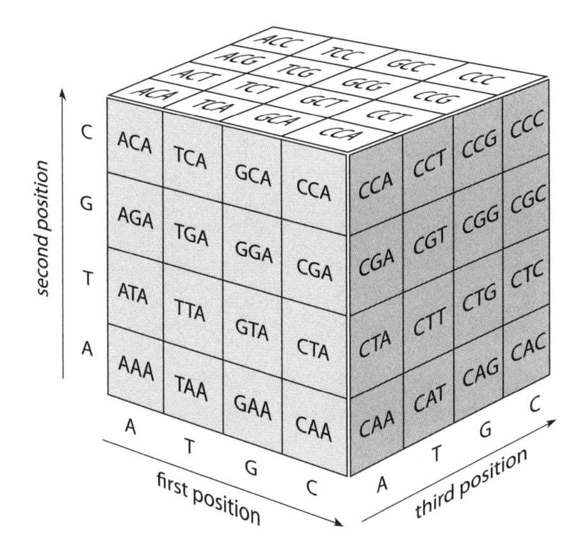

Figure 7. Possible sequences for one strand of a DNA molecule three base pairs long. Only three outer surfaces of the cube are shown (there are more possibilities inside).

represent a different combination of four bases on one strand, such as AGCT or AGGT. Then each genome corresponds to one out of 128 possible locations in genetic space.

What about a genome of one hundred base pairs? This would require a hypercube with one hundred dimensions to represent all the possibilities. We cannot visualize such an object—even Einstein struggled to think in four dimensions, let alone one hundred. Talking about many dimensions is just a convenient way of saying that we can vary many features independently from each other. And because of the principle of combinatorial richness, the more dimensions we have, the vastly greater the number of possibilities. For a one hundred-dimensional hypercube with four bases along each axis, there are $4^{100}/2$ different genomes, which is about 10^{60} (a one with sixty zeros after it). This is an astronomically large number, much larger than the number of organisms that have ever inhabited earth. Simply by repeating the same unit one hundred times and allowing each unit to vary among four possible states, we have created an enormous space of possibilities. This is the principle of combinatorial richness in action, where the variable components here are molecular modules, or bases in the DNA.

Like the mind-boggling range of possibilities for DNA variation, there is a correspondingly enormous potential for protein variation. For a protein that is one hundred amino acids long, the range of possibilities can also be represented with a one hundred-dimensional hypercube, but in this case there are twenty divisions along each axis instead of four (corresponding to the twenty different amino acids). This gives 20^{100} (about 10^{130}) possibilities, more than the number of atoms in the observable universe, which is a mere 10^{78}. Again, it is the principle of combinatorial richness that creates this enormous space of possible proteins.

These one hundred-dimensional hyperspaces are just the tip of the iceberg. The human genome contains about twenty-five thousand protein coding genes, each of which may be several thousands of bases long. If we consider the range of possibilities with twenty-five thousand gene dimensions, each of which has a large hypercube of possibilities, we are confronted with a colossal hyper-hypercube. The same applies to the twenty-five thousand proteins that are encoded by these genes. The range of combinations of amino acids and proteins is truly mind boggling.

When the human genome was first sequenced, many scientists expected it to contain more genes than were found in other creatures. After all, we seem to be much more complex than any other organism. Initial estimates put the number of human genes at around seventy thousand, more than for other organisms. But as knowledge about the genome improved, the number was continually revised downwards. Recent estimates put the number of protein coding genes in the human genome at around twenty-five thousand, about the same number of genes as in a small weed called *Arabidopsis*. At first this sounds surprising because humans seem far more complicated than weeds. But even twenty-five thousand genes contain vastly more possibilities than we can conceive of. Life is interesting because of the enormous combinatorial richness of organisms and their environments.

Wandering Clouds

To get a sense of what so many genetic possibilities mean for evolution, it helps to visualize the process of evolutionary change. Because it is not easy to concieve of large hyperspaces, an alternative is to imagine a vast three-dimensional space. Think of all the genetic possibilities as one enormous space, with each point in this space corresponding to one particular genome sequence. Then the genome of any individual would correspond to one position in this vast genetic space of possibilities. Your genome sequence, for example, would rest at one particular location in genetic space. Each person has a slightly different genome sequence because of mutations and the continual shuffling and recombination of chromosomes. Consequently, the human population as a whole would correspond to a cluster of points, or a cloud in genetic space. This cloud would be very small in relation to the entire space because only a tiny fraction of potential genetic variation would be exhibited by any one species. Because there are many species on the planet, there would be many clouds like this, each occupying a tiny part of genetic space.

We can now think of evolution as shifts in the position of population clouds, corresponding to changes in the genetic composition of species. During evolution, each population cloud effectively explores particular regions of genetic space, moving along in various directions. As there are many species, there are many clouds moving through genetic space at any one time. Even so, because genetic space is so vast, only the tiniest fraction of this space has ever been explored by life on earth.

To gain a more concrete idea of what this means, let's look at how a single protein might evolve. The ability of plants to fix carbon dioxide from the atmosphere depends on a protein called ribulose bisphosphate carboxylase, or Rubisco for short. This protein locks onto molecules of carbon dioxide from the atmosphere and encourages them to fix onto other carbon molecules. At first sight, Rubisco seems to be effective at fixing carbon dioxide—it can fix about three molecules of carbon every second, which sounds quite fast. But this is actually rather slow for a protein; other proteins can complete hundreds or thousands of reactions per second. So Rubisco is rather sluggish. Slowness is not its only defect; the protein also frequently makes errors. Sometimes, it fixes oxygen instead of fixing carbon dioxide. Fixing oxygen is wasteful for the plant because it eventually leads to the release of carbon dioxide, the opposite of carbon fixation. As much as one third-of the carbon that is fixed in plants is lost again because of this molecular error. The problem is that the shape of Rubisco is just not specific enough to always discriminate between carbon dioxide and oxygen molecules.

The problem is particularly acute for plants living in hot, dry conditions, where the opening of leaf pores must be restricted to prevent water loss. Closed pores cause the levels of carbon dioxide in the leaf to fall, because carbon is being fixed more quickly than it can be replaced from the atmosphere. Rubisco then makes more errors: there are fewer carbon dioxide molecules relative to oxygen molecules, so oxygen instead of carbon dioxide is bound more often. A solution to this problem has evolved in several plant species. In these plants, Rubisco is only produced in specialized cells of the leaf. The other leaf cells pump the low levels of carbon dioxide in the leaf toward these special cells, raising their local carbon dioxide levels. This way, even though the overall concentration of carbon dioxide in the leaf is low, its level in the specialized cells with Rubisco protein is high, reducing the chance of an error.

On the face of it, Rubisco is far from optimal. It does a reasonable job of fixing carbon but perhaps somewhere in the space of possible proteins there is another protein that would fix carbon dioxide more quickly and effectively. The process of evolution may never have led to this protein version because only a tiny fraction of the possibilities have been explored. But even though Rubisco may not be optimal, natural selection has led to several ways of getting around its limitations. One is by producing abundant quantities of the protein so that there is enough to fix the required carbon. In fact, plants produce so much Rubisco that it is thought to be the most abundant protein on Earth. The second method, as we have seen, is to pump carbon dioxide into specialized cells to raise its local concentration. So rather than being blocked by the limitations of Rubisco, other routes and devices have been found to get around the problem.

You might regard the solutions of producing more protein or of pumping carbon dioxide within the leaf as ugly hacks that reveal natural selection's incompetence in finding an optimal protein. Or you could equally well view the solutions as beautiful adaptations that work with the constraints of the system. Either way, what matters is that problems can be circumvented in all sorts of ways because genetic space is so vast and has so many directions available for exploration.

This flexibility means that problems are often solved in diverse ways as evolution proceeds. Consider the evolution of flight. Feathers were a key feature that enabled flight to evolve in birds. So, it might seem that not having feathers would have been a block to the evolution of flight in other groups, like mammals. Yet flight has evolved in mammals, like bats, by the problem being solved in a different way. Instead of feathers their wings comprise extended flaps of skin, stretched between their long, thin fingers. Similarly, compare different ways of chirping. Birds chirp through vibration of membranes in their air passages, while crickets rub their wings together. Multiple genetic options enable multiple solutions.

A vast genetic space of possibilities has two consequences. The first is that only a tiny part of that space can ever be visited by our wandering population clouds. The second is that the high dimensionality of the space means that there are many directions for the clouds to move, providing enormous flexibility. What natural selection loses by not being able to cover all possibilities, it gains in the number of directions it can take.

Human artistic creations may be viewed in a similar way. One of the limitations of a painting is that it cannot show movement directly. However, artists have played with various devices to convey movement within these constraints. A good example is Michelangelo's painting on the ceiling of the Sistine Chapel of God creating Adam (fig. 8). The leftward movement of God is conveyed by the billowing drapery and his windswept hair, together with his outstretched arm and the suggestive gap at the fingertip that is asking to be closed.

A different approach to conveying motion was taken more than 350 years later by the Impressionist Claude Monet. In *The Rue Montorgueil*, Monet painted numerous flags in a jumbled and imprecise manner (fig. 9, plate 3). Through this lack of definition he achieved a remarkable sense of movement and vitality. Our eyes seem unable to pause on any one thing, and consequently the flags appear to flutter before us. After Michelangelo, you might think that it was only by twisting the body further, or ruffling the drapery more, that additional movement could be represented in a painting. Indeed, many followers of Michelangelo tried to capture movement in precisely this way, displaying ever more elaborate bodily contortions. Yet by taking a different approach, Monet conveyed motion in an entirely fresh way.

Painting involves the application of patches of color to a fixed surface. The number of possible color combinations—the set of all possible paintings—is

Figure 8. *God Creating Adam*, Michelangelo Buonarroti, 1510.

unimaginably large. Artists can explore only a tiny fraction of this space of possibilities, yet within this constraint they are able achieve remarkable results because there are so many avenues that can be explored.

Principle of Recurrence

We have now covered six principles of evolution. In the previous chapter we saw how the four principles of population variation, persistence, reinforcement, and competition interact to provide the foundations of natural selection. In this chapter two further principles have emerged. One is the principle of cooperation—proximity allows fates to become intertwined, so natural selection leads to the formation of collaborative units such as genes, proteins, cells, and individuals. These cooperative units are both an outcome and ingredient of natural selection. The other principle is combinatorial richness which leads to a varied world and a vast space of genetic possibilities.

Figure 9. *The Rue Montorgueil, Paris, Celebration of June 30, 1878*, Claude Monet, 1878. See plate 3.

A further issue has been lurking in the background, however. Throughout this account I have been taking particular contexts for granted. In the example of natural selection given in the previous chapter, I assumed a population of apple trees interacted with a feature of its environment, a population of bears. I also assumed that apple trees had a particular genome from which variations could arise, such as different degrees of apple sweetness. Our six principles of evolution are always operating within a given context like this. How does this context arise?

The answer involves the way that natural selection feeds back on itself. It might seem that once an organism has become better adapted to its environment, natural selection would become less intense. But this does not happen. To see why, imagine you are competing in a car race and invent a gadget that makes your car go faster than others. Initially you have an advantage in winning, but once everyone learns about the gadget and it becomes incorporated in all cars, your advantage is lost because all cars now go faster. What was originally an advantage is now simply a standard feature of all cars, a new context upon which further improvements can be made. Rather than reducing competition, the spread of the gadget has simply improved everyone's performance, pushing inventors to come up with even better gadgets so that their cars can win the race.

Similarly, once an adaptation gets fixed through natural selection it is shared by all individuals in the population. Competition is even fiercer than it was before because now all individuals in the population are better able to reproduce and compete for limited resources. Natural selection does not weaken as it promotes the spread of adaptations, but spurs itself on as each adaptation shifts the context and makes competition even more intense. Our imaginary population cloud propels itself on through genetic space, with each new adaptation creating the conditions that help drive the next. Just as a car race is not about attaining a particular speed but about who goes faster relative to the others; natural selection is not concerned with reaching a certain level of adaptation but operates in a relational way, continually favoring gene versions that reproduce more effectively in relation to others in the population. There is no "time out" for natural selection because the competition never stops, but is intensified through the adaptations already brought about.

I call this relentless repetition of a process, always spurred on by what went before, the principle of recurrence. Without recurrence, organisms would represent just a crude stab at capturing one environmental variable, the outcome of a single step of natural selection. Instead, we have a continuing process of adaptation building on adaptation, driven on by the ever shifting standards it creates. Consequently, our population clouds rarely stay still, but instead keep moving as new contexts create further challenges.

Evolution is continually propelled by the past, always modifying its own context and creating the conditions for new steps. This dependence on the past makes the process highly constrained by history. The Rubisco protein, for example, first evolved billions of years ago when there was little oxygen in the atmosphere, so its inability to discriminate between oxygen and carbon dioxide molecules would not then have been of great importance for plants. Incorporating oxygen instead of carbon dioxide only became a significant problem later, when oxygen levels had increased in the atmosphere through plant photosynthesis. But by then, it might have become difficult for the structure of Rubisco to be modified to solve this problem, due to the structure already having certain restrictive features in place. Similarly, the evolution of flight in birds and bats was constrained by the particular arrangement of bones in the forelimbs of their ancestors. In the case of birds, wings correspond to entire arms; while in bats, they correspond to splayed out hands. However, the basic arrangement of bones in the forelimbs is preserved in both cases and reflects the ancestral arrangement. The price of always building on what went before is the constraint caused by earlier systems. This constraint need not prevent further change because, as we have seen, combinatorial richness presents many evolutionary options. But the constraint of history often results in evolutionary paths that are highly convoluted and idiosyncratic.

The same is true for our artistic analogy. Monet's innovations depended on reactions to a string of previous achievements in painting, including the advances of Michelangelo and others in the Renaissance. If Monet had been born four hundred years earlier, he could not have produced paintings in the Impressionist style. Similarly, the discoveries in the Renaissance depended on artistic developments in the Byzantine period and classical antiquity. Even though art is based on earlier achievements, the story of art is not a straightforward march. In some respects the Impressionist style may seem cruder than the realism of the Renaissance, and Byzantine paintings with their elongated and flattened figures may seem more primitive than the paintings of classical antiquity that preceded them. As with biological evolution, art follows circuitous paths because it is continually building on and reacting to what went before, shifting its own context.

The Origin of Species

So far we have looked at how recurrence operates within a single population. By continually building on what went before, by one context setting the scene for the next, our population cloud moves through genetic space, exploring different regions. But evolution does not involve just a single population. There are many species, each of which can be considered a population of interbreeding individuals. There are therefore numerous population clouds, or species, wandering about in various directions. Moreover, these species may interact with one another, as we saw with apples and bears. These interactions provide an even broader context for evolution that involves entire ecosystems. How does this broader context arise?

To answer this question, we first need to know what keeps a population together. What stops a population cloud from dispersing into the vast genetic space around it? The answer has to do with sex. If an individual is very different from the rest of its population, it might be unable to mate effectively with a partner. And even if it does mate, the offspring could be such a hotchpotch of different traits that they would not easily survive. Consequently, the continual interbreeding of individuals, combined with natural selection, ensures that the variation of a population is restrained within certain limits. This restraint keeps members of the same interbreeding population close to each other in genetic space. Regular sex provides the cohesion, the genetic glue that keeps the population together and defines it as a unit or species.

But this raises a problem, one that caused Darwin sleepless nights. If sexual reproduction holds species together, why is it that we have so many different populations on earth, millions of different species? If we imagine that all species descend from a common ancestor, then at some point this ancestral species must have split into two or more species. Presumably, successive rounds of splitting like this have resulted in the many species of today. But how is the sexual glue that holds species together ever broken down so that one species becomes two?

It is difficult to see how natural selection alone could drive this process of speciation. Natural selection is a process that promotes an increased reproductive ability, yet a key feature of two species is their inability to breed with each other. Even if they successfully mate, two different species usually produce infertile offspring, as when a donkey and horse breed to produce a sterile mule. It is not easy to see how infertility, the opposite of reproductive success, could be selected for. Having argued that sexual reproduction holds our population clouds together, it is difficult to see how clouds ever split apart, how one species is cleaved into two.

We encounter the same problem with languages. One of the most bizarre features of humankind is that it converses in so many different languages. For the sake of communication, it would make more sense for everyone to speak the same tongue. The Bible had to go to considerable lengths to explain why the Creator might have imposed linguistic diversity. According to the book of Genesis, everyone originally spoke the same language, and diverse tongues arose later as a punishment meted out by God to the people of Babel. This took place when these people decided to celebrate their achievements by building a great tower together (fig. 10). Such an act of vanity brought about God's disapproval, and he punished the inhabitants of Babel by making them speak different tongues, which destroyed their ability to work together. They then dispersed, giving rise to the many languages spoken by diverse peoples of today. It is a strange story designed to account for a strange choice—if you were creating humans, it would make more sense to have everyone speak the same language rather than different ones.

We can come up with a more mundane explanation of how linguistic diversity arose. Consider European languages such as English, French, or Italian. These languages are so distinct that speakers of one cannot readily understand those of another. Yet all of these languages share common roots—they all trace back to a common ancestral tongue, thought to be spoken by the early founding tribes of Europe. If you imagine all conceivable languages as an enormous space of possibilities, then it is as if an ancestral cloud in this linguistic space gradually gave rise to many different clouds, constituting the various languages of today. Throughout this process, each

Figure 10. *The Tower of Babel*, Pieter Bruegel the Elder, 1563.

linguistic cloud has remained coherent—individuals sharing the same language have continually been able to converse with each other—yet over time, the clouds have moved so far apart that conversations between individuals from different clouds have become difficult.

Geographical barriers have probably played an important part in linguistic cloud splitting. As long as people are regularly conversing with each other, there is a strong tendency for them to speak the same language. However, once populations become separated by mountains or a sea, languages can begin to diverge because there is no longer any need to maintain communication between the two groups. It is not that differences in the physical environment drive the changes—speaking Italian, rather than English, does not allow you to deal more effectively with the climate in Italy. Instead, languages tend to drift apart through the accumulation of small changes that make little difference one way or another on their own. As each generation learns how to communicate, chance variations in words and their pronunciations may be introduced, and then spread through the region. Some words are established at the expense of others because there are limits to how many words a language can sustain. For example, if there are two ways of saying "apple," then over time we might expect one of them to predominate. It is not that one way of saying apple is intrinsically better than the other, only that there are limitations to the number of words we can retain, so one version is likely to supersede eventually.

Each change on its own does not disrupt comprehension, but by accumulating many such changes, languages may evolve to be very different from each other; so different, that the speakers of one language may no longer understand those speaking another. In our imagined linguistic space, each language forms a cloud, which is kept small because there are only minor linguistic variations among people that regularly converse with each other. But when barriers to conversation arise between people, most often because they are separated from each other geographically, linguistic clouds may split and drift far away from one another over time. Of course, this account of linguistic divergence is greatly simplified, but it illustrates how clouds can drift apart even if there is nothing actively driving them to diverge.

Equivalent divergence may arise in the case of biological evolution. Physical barriers are common on earth because matter aggregates at many different scales, producing an uneven world that is full of boundaries. Rivers, seas, or mountains may divide one region from another. On a smaller scale, ponds may be separated by a tract of land. Populations are then restricted in their movements at one scale or another. Occasionally, a few individuals from a population may overcome a barrier and invade a new location, swimming to an island or traveling from one pond to another, creating a new colony. The result is two spatially separate populations. And because sexual reproduction is a cooperative act that depends on physical contact—the union of the sex cells from each parent—there is little or no genetic exchange between the two populations.

In the absence of contact, our two population clouds may start to wander off in different directions in genetic space. This divergence may be driven partly by environmental differences between the two locations that lead to distinct adaptations through natural selection. But even without major environmental differences, we might expect the clouds to drift apart. One reason has to do with trade-offs: being better in one respect is often at the expense of being worse in another. Having hooves, for example, may mean you are better at running along the ground, but can make you worse at climbing trees. One advantage is traded for another. For some evolutionary changes, the benefits gained by adapting in one respect may be more or less compensated for by losses in another respect. The advantage of being able to run faster may be balanced by the disadvantage of not being able to climb trees. In such cases, an evolutionary change may lead to no major difference in an organism's overall ability to survive and reproduce, even though it may modify the way the organism functions. In evolutionary terms, the change is almost neutral, equivalent in our linguistic analogy to substituting one way of saying apple for another equally good way. The higher the dimensionality of our system, the more likely it is that such trade-offs arise because there are more possible options to explore. Given the many dimensions of genetic space and the environment, we might expect that there are many paths of this kind along which our population clouds may move. Trade-offs are common because combinatorial richness ensures there are many equivalent directions that life can take.

If a population is divided by a physical barrier, like a mountain range, the separate populations are likely to start moving away from each other in genetic space and operate in different ways, just as separate languages diverge over time. And in moving apart like this, the populations may also cease to be sexually compatible, just as communication between those speaking diverging languages becomes increasingly difficult. The sexual incompatibility is not driven by natural selection but is an indirect consequence of the reproductive barrier between our clouds. To put it another way, like communication, sexual compatibility between individuals in a population has to be actively maintained through regular exchanges; once these exchanges stop, incompatibilities may start to accumulate. I have given one example of how cloud splitting (the division of one species into two) may take place following geographic isolation, but this should not be taken as the only mechanism: there are many other possible scenarios involving various degrees and forms of separation.

Species and Ecosystems

We have seen how a species can be split in two following separation by a physical barrier. But how has evolution led to the formation of multiple species in the same location? This involves many clouds in genetic space coexisting in the same region of physical space. To answer this question, let's look at the possible outcomes when some contact is reestablished between populations after a period of separation.

Suppose a few individuals from one of our populations happen to cross the separating barrier and arrive in the area inhabited by the other population. If the newcomers are still able to breed with those in the population they encounter, they may gradually become fully integrated into it over several generations. The genetic cloud of the newcomers becomes absorbed into that of the host population.

Another possibility occurs when the populations have diverged so much that interbreeding is no longer possible. In this case, the newcomers may form a separate cohort that coexists with the larger population. If the new cohort lives in a very similar way to its neighbors and freely mixes with them, competition for the same limited resources is likely to ensue. This may eventually lead to one of the populations going extinct, due either to it being less able to compete or to the laws of chance (if both populations compete equally well). However, if the newcomers function differently than their neighbors, they may be able to exploit their environment in distinctive ways, and thus occupy a slightly different niche. Many possible niches are likely to be available because of the combinatorial richness of physical environments and the organisms within them. By occupying slightly different niches, the populations could coexist more stably because each could be sustained by resources that are to some extent distinctive. And because there are benefits to not competing for the same resources, we might expect natural selection within each population to increase specialization, driving the populations further apart over time. In other words, the two genetic clouds would separate even further, as if repelling each other. The result is that the two population clouds become increasingly distant in genetic space, even though they occupy the same region of physical space.

We have ended with a small ecosystem comprising two species, each exploiting a different niche. I should emphasize that these niches are not absolute features of the environment; they arise through interactions between our two populations. If there was only one population, the notion of a niche would be unnecessary. It is only when we consider populations interacting and specializing that separate niches start to become defined. A niche is a relational concept, a feature that evolves together with species rather than existing independently.

With populations continually splitting and coming back into to contact, multiple species can evolve with each occupying a distinctive niche in an ecosystem. The result is many organisms living together, each capturing particular aspects of their environment. Of course there is a limit to this cloud splitting because ecosystems occupy finite regions and cannot sustain an infinite number of species. As well as new species forming, others may go extinct. The ecosystems of our world reflect this balance between the production and extinction of species. Because of human activity, the rate of species extinction is currently much higher than the rate at which species are originating, so the number of species is now rapidly declining overall.

With multiple species in mature ecosystems, many complex scenarios are possible for how our population clouds interact, move, and split apart. It is now possible for trade-offs to arise through interactions among species. The sugars that make an apple sweeter to a bear may also make it more attractive to fungi and bacteria: a gain in seed dispersal is traded for increased susceptibility to disease. Natural selection in the apple tree population may act to increase disease resistance, but at the same time natural selection in the populations of fungi and bacteria may act to increase their virulence. Populations do not evolve in isolation; they are continually evolving in relation to each other. A continual realignment and jiggling of species and niches takes place within every ecosystem. As each population cloud moves, it influences other clouds, altering their context and nudging them in new directions. Genetic space is teeming with population clouds driving themselves on while also pushing each other in new directions. This is the principle of recurrence operating at the level of many interacting species.

A Recipe for Evolution

In the previous chapter we saw how a double feedback loop between reinforcement and competition lies at the heart of natural selection. These two interconnected loops are fed by a balance of population variation and persistence, leading to an increased ability to survive and reproduce.

But this view of evolution takes certain things for granted. It assumes that we have a population of integrated spatial packets called individuals. We have seen that individuality depends on cooperation, which is both an outcome and ingredient of natural selection. By bringing multiple components together, cooperation also allows components to be combined in a large number of ways, giving many possibilities. This principle of combinatorial richness is what makes the world and its organisms so pregnant with possibilities.

The interplay between these six principles—population variation, persistence, reinforcement, competition, cooperation, and combinatorial richness—provides the foundation for evolution. But the reason that it generates such a rich and varied living world has to do with recurrence, our seventh principle. Every adaptation shifts the goal posts, setting the scene for further adaptations. Rather than competition diminishing as adaptations spread in a population, it increases as the stakes are raised, spurring the population on to meet higher standards. And as populations split, diverge, and come back together, the same process takes place across multiple species, each one vying with the others for a sustainable place in a complex ecosystem. The context for evolution is not fixed; it is continually shifting and being redefined by the evolutionary process itself.

As evolution continually shifts and redefines its own context, you might wonder what the very first evolutionary context was. Life originated around 3.8 billion years ago, but the precise circumstances that led to the first living forms are not known. It is unlikely, however, that there was a clearly defined beginning because, as we have seen, evolution involves numerous feedback loops and the nature of loops is that they do not have clear starts. Crude molecular assemblies with an elementary ability to replicate could have been a primitive form of cooperation and persistence. These assemblies could then have been reinforced through replication, eventually bringing about competition for limited resources. Variations arising during the process of reproduction would then be subject to further reinforcement and competition, driving higher levels of cooperation. As cooperation increased, the components brought together could be combined in various ways, leading to greater combinatorial richness and thus a greater diversity of possibilities. And as this process recurred, the first forms that we might recognize as living beings would have emerged. The context of early life could therefore have arisen through a series of feedback loops and interactions, rather than in one clear step.

Whatever the events were that led to the first stirrings of life, the richness of life is not to be found in these early events any more than the richness of paintings is to be found in the first human scribbles in the sand. Rather, life's splendor lies in the way these initial stirrings have been dramatically transformed. As the earliest population clouds emerged, they spurred themselves along many journeys through genetic space. Some split to create new clouds, while others petered out and went extinct, but their journeys through the vast space of genetic possibilities were relentless. It is through these voyages that the rich tapestry of living patterns and forms arose.

Our seven principles—population variation, persistence, reinforcement, competition, cooperation, combinatorial richness, and recurrence—and their interactions provide the driving force for these journeys, leading to the remarkable variety of organisms we see today. I have called this collection of seven principles and how they work together *life's creative recipe*. Life's creative recipe is grounded in the physical world. It relies on the action of physical forces and the limitations of a finite world. This is because life is a manifestation of matter. But it is a very peculiar manifestation, one that arose when the ingredients of life's creative recipe first came together several billion years ago.

Evolution is a process of transformation, one that has made several other types of transformation possible—the development of eggs into adults, the conversion of newborns into sophisticated adults, and the transformation of tribal communities into complex civilizations. How do these latter transformations compare with evolution? To answer this question we must look at how these other types of transformation work. We will begin with development, the first of these transformations that was to emerge through evolution.

THREE

Conversations of an Embryo

A POSTMAN KNOCKS at your door and hands you an envelope containing an invitation. It reads: "We have the pleasure of inviting you to attend a tea party in honor of Alan Turing." You have a friend staying with you who prides himself on his general knowledge, so you show him the invitation. "Ah yes, Turing. I think he was one of the guys who invented the computer. He also helped to break the secret code used by Germany during the Second World War—the Enigma code I think it was called. He committed suicide when quite young in the 1950s—bit an apple dipped in cyanide. Looks like an intriguing invitation to me. I would go if I were you."

You arrive at a large building in the middle of Manchester. After handing in your invitation at the door you are ushered into a large room and given a cup of tea. There are lots of people milling about on their own, but eventually someone approaches.

> "Hello, I'm Charles. Do you know what this is about?"
> "Sorry, I have no idea. I just got this invitation out of the blue."

Someone else overhears and joins in.

> "Oh, you as well? Yes, this is very odd. I'm a mathematician and I work in the same department as Turing did when he was here in Manchester, but I have no idea what is going on today."

A few more people join in the conversation but now your group is getting rather large, and you can't hear what everyone is saying over the general noise in the room. So, you gradually edge away from the group and stand on your own again. By now lots of groups have formed with people busily chatting away. You feel rather conspicuous by yourself, so eventually you join another group of people and enter into their conversation. The tea party continues in this way with lots of small clusters of people chatting together. Even though the individuals within each group change over time, the clustering pattern persists. Turing would have been pleased.

Throughout his life, Turing was interested in discovering simple mathematical relationships that lie behind patterns. His early work showed that even our ability to solve problems is a type of pattern that can be expressed mathematically. This idea led to the notion of problem-solving machines, better known as computers. He also worked on how to unscramble the pattern of symbols in coded messages during the war. Toward the end of his career, Turing turned his mind to the patterns that arise through biological development. By *development*, I mean the process by which a single cell, the fertilized egg, is able to transform into a complex, multicellular organism. Typical of Turing, he simplified the problem to arrive at the essence of this transformation.

Look at figure 11 which shows a spaced out pattern of spots. This pattern was not arrived at by carefully drawing the spots by hand, but was generated automatically by applying some simple mathematical rules. These rules were first formulated by Alan Turing in 1952 to show how molecules could organize themselves. The level of gray in the diagram refers to the concentrations of a molecule, with black showing regions of high concentration, where the molecule is abundant, and white areas showing where the concentration is lowest. You can see that the molecules are clustered together in regions, forming spaced out areas of high concentration, rather like the clustered conversations at the tea party. No one told anyone at the tea party where to form conversational groups; the groups emerged spontaneously by people interacting. Similarly, the regions of high molecular concentration in figure 11 arise from an initial, uniform distribution of molecules and their subsequent interaction. When defining the rules of interaction, Turing highlighted some key principles by which patterns can be transformed from initially simple distributions into more complicated arrangements. Because these rules involved molecules, Turing thought they might explain how an apparently simple initial structure, the fertilized egg, could gradually turn itself into a more

Figure 11. Pattern obtained by applying Turing's rules.

complex adult. Let's look more closely at the main components of Turing's system. In what follows, I use some of the same terms used for describing evolution in order to highlight fundamental elements common to both.

Turing's Principles

If you place a tea bag in a cup of cold water, the water starts to turn brown as the tea dissolves and spreads from the tea bag to the rest of the cup. This process of diffusion involves a smooth progression of the brown color. Yet if you could zoom in to observe the individual water and tea molecules, instead of a smooth movement you would see frenzy and chaos. The molecules continually jiggle around and collide with each other in a random manner. This behavior is a consequence of the physical rule that things close to one another tend to interact most strongly; when two molecules in solution move too close, they usually repel or bounce off each other. We have two levels of behavior: at the scale of the tea cup we see a smooth progression of brown color, while at the scale of molecules we see frenzy and chaos.

These two levels of behavior can be related through the ideas of statistics. As we saw in chapter 1, statistics involves the notions of population and variation. The population in this case is the collection of water and tea molecules in the cup. Variation involves the random motions and collisions of the individual molecules. Even though each molecule follows a random trajectory, the tea molecules tend on average to move away from their source, the tea bag. This is because there is nothing to keep the tea molecules near the tea bag, so they are likely to move away from it after a while, even if they move in random directions between collisions. For a similar reason, a drunkard taking many random steps after leaving a house is unlikely to remain in the vicinity of the house. The overall result of the random movements of our numerous tea molecules is their gradual spread through the cup. Because the number of molecules and collisions is so vast, at the level of the population as a whole, we see a smooth progression of the brown color. From the chaos of individual molecular movements emerges regular diffusion. As with other statistical processes, diffusion is not a property of a single event taken in isolation (individual molecular movements between collisions are in random directions), but the collective outcome of a population.

There is an average distance over which the tea molecules tend to diffuse in a given interval of time. This means that diffusion can provide a type of ruler, a measure of distance. But unlike the rulers we normally use for measuring, this is a statistical ruler that operates according to the population behavior of molecules.

In addition to diffusion, random molecular collisions can lead to another type of event—a chemical reaction. Two molecules may come together and join to form a new molecule, just as two people may bump into each other and start a conversation. Because chemical reactions depend on random collisions, the rate at which they occur depends on the abundance, or concentration, of the molecules involved. This dependence is sometimes called the law of mass action. It arises as the average effect of many random collisions within a population of molecules, and is therefore another statistical property. These two forms of statistical regularity—diffusion and chemical reaction rates—provide the foundations for Turing's patterning system. For this reason, the process he described is sometimes referred to as a reaction-diffusion system.

Too much variation can be a problem if we want to generate a pattern—our tea party would fail to form groups if everyone was madly rushing around. So, in addition to movement in our population of molecules, we need an element of stability. In the case of Turing's system, the products of the chemical reactions have to be stable enough to persist for some time, and the molecules should not be stirred up or mix with each other too rapidly. There needs to be sufficient variation (random molecular movement) to allow patterns to start forming, but not so much that everything dissipates. As well as population variation, we need persistence.

A key feature of a tea party is that once two people start chatting, others like to join in. Conversation encourages more conversation, and thus boosts itself locally. Turing proposed an equivalent type of reinforcement for his molecular system. As well as participating in reactions, some molecules can catalyze or promote reactions. Turing looked at what would happen if the product of the reaction was itself a molecule that acts as a catalyst. In other words, suppose we have a catalyst that enhances a reaction, and the product of the reaction is more of the catalyst molecule. The catalyst molecule would then boost its own production, essentially promoting itself. If the concentration of this molecule is slightly higher in one area, even more of it tends to be produced there. It is a system that promotes disparities, amplifying differences rather than smoothing things out.

If all we have is reinforcement or self-promotion, the system as a whole would simply boost itself to ever higher molecular concentrations, rather than producing an ordered pattern. This brings us to our next principle—competition. In the case of the tea party, the tendency for people to get together does not continue unabated. As a group grows larger, it becomes more difficult to hear what people are saying above the noise of other conversations, so people tend to leave groups after they reach a certain size. Each group becomes a victim of its own success. That is why we end up with many separate conversational groups rather than one massive collection of people. In this example, the self-limiting aspect of group size arises because people compete to hear what is being said. Another source of competition is the struggle to speak—as a group gets bigger, you may find less opportunity to contribute and may choose to leave.

Turing also invoked a form of competition in the generation of molecular patterns. One way this competition can arise is through a natural limitation on the chemical reaction. Production of a molecule through a reaction involves consumption of other molecules, called substrates, that are present in the solution. Every time a self-promoting molecule is formed, substrate molecules are used up. But the supply of substrates is not infinite, so as the process continues the substrate starts to run out. The self-promoting molecule becomes a victim of its own success—as more of it is made, less substrate is available for its synthesis. The self-promoting molecules start to compete with each other for the limited substrate that remains. This competition becomes more intense as more substrate is consumed, slowing down the increase in the self-promoting molecule even further.

The process can be summarized by two interconnected feedback loops (fig. 12). The self-promoting molecule forms a positive feedback loop because it boosts its own production (reinforcement). The negative feedback loop indicates the constraining effect of substrate limitations on the system (competition). These loops operate in the context of diffusion: both the self-promoting and substrate molecules continually jiggle around, allowing them to spread and undergo chemical reactions. Turing showed that depending on how fast the self-promoting and substrate molecules are able to diffuse, this situation can sometimes lead to spaced out regions of high concentration, as in figure 11. This can happen even if we begin with a uniform distribution of molecules. As with conversations at the tea party, self-promotion can lead to local boosting, while competition breaks things up and helps to space them out. Turing had shown how some simple rules could lead to the transformation of an initially uniform distribution into a more complex pattern of spots.

Another way of achieving the same type of result involves a slightly different form of competition. Suppose that the self-promoting molecule not only catalyzes its own production but also that of another molecule, which we call an inhibitor. The effect of the inhibitor molecule is to interfere with the self-promoting molecule, preventing it from catalyzing its own production. Again, we have two feedback loops—one positive and

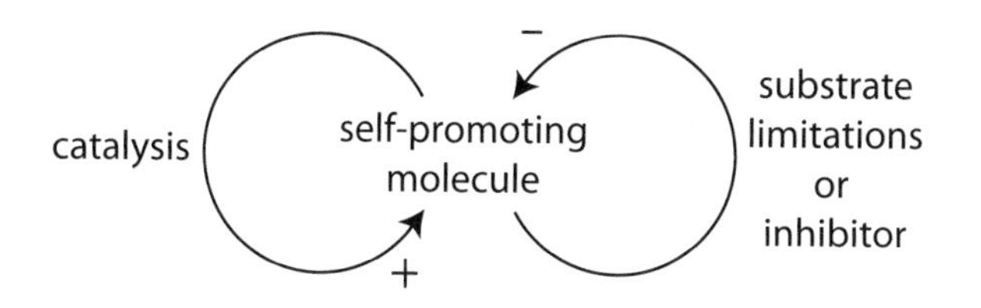

Figure 12. Interplay between reinforcement (positive loop) and competition (negative loop) for a Turing system.

one negative—but the negative loop now represents an inhibitor rather than the limiting substrate. As the self-promoting molecules start to increase, they catalyze production of more inhibitor molecules, which in turn counteract the rise of the self-promoting molecules. The self-promoting molecule has again become a victim of its own success because as more of it is produced, more of its inhibitor is also generated. Once more, the result can be a spotty pattern of high concentrations.

In this case the limitation in the rise of the self-promoting molecule is brought about by production of an inhibitor rather than competition for a limited amount of substrate. Nevertheless, the fundamental principle of a negative feedback loop driven by a self-promoting entity is very similar. I use *competition* as a general term to describe both forms of the negative feedback loop—as limitations brought about by inhibition or as limiting amounts of a component.

At the root of these patterning systems we find some formal similarities with what we encountered for natural selection in chapter 1. In both cases, we find a double feedback loop involving reinforcement and competition, fed by a balance between population variation and persistence. Unlike natural selection, though, Turing's system does not involve populations of organisms reproducing over generations, but instead involves populations of molecules interacting over a much shorter period of time. It involves a similar set of principles as those of natural selection but they appear under a new guise. The net result is not the evolution of organisms, but the transformation of patterns within an individual organism.

At the time Turing proposed his ideas, little was known about the molecular mechanisms that underlie patterning. The way genes work was also poorly understood—even the structure of DNA had yet to be discovered.

His scheme was rather speculative with very little direct evidence for or against. In the past few decades, however, we have gained a much better understanding of how development actually takes place. Although these more recent findings are sometimes described in different terms from those Turing used, we will see that they involve the same sort of principles that he invoked.

Patterning a Cell

To convey the picture that has emerged from more recent discoveries, let's start with a case of patterning from the unicellular world. Every time an individual bacterium of the species *Escherichia coli* reproduces, there is a patterning event—a partition forms in the middle of the cell, dividing it in two. How does *E. coli* manage to draw a line down its center? The positioning of the partition depends on an interaction between two proteins, called MinD and MinE. These proteins police the two ends of the cell, preventing a partition from forming there and ensuring that it can only form in the middle. If you were to look at MinD and MinE over a period of time you would see a high concentration of them at one end of the cell, and then a few minutes later, a high concentration at the other end (fig. 13). They behave like policemen on a regular beat, moving back and forth to ensure that no partition is allowed to form near the ends of the cell. What underlies this curious behavior of the Min proteins?

We begin with MinD. If you were to look closely at where the MinD protein is found, you would see that in addition to freely diffusing within the cell, it can also bind to the cell membrane. Once a MinD protein is attached to the membrane, it can stay bound and persist there for some time. This binding to the membrane also encourages other molecules of MinD to bind nearby. In other words, MinD reinforces the binding of more of its kind. You might expect that eventually all of the MinD molecules would become attached to the membrane. However, this doesn't happen due to the competing effect of MinE, the other protein in our partnership. MinE can bind to molecules of MinD on the membrane, but when MinE attaches, MinD then falls off the membrane and moves back into the cell's interior. This means that MinD becomes a victim of its own success; as more binds to the membrane, more MinE is attracted there, and more MinD then falls off. The relationship can be summarized with

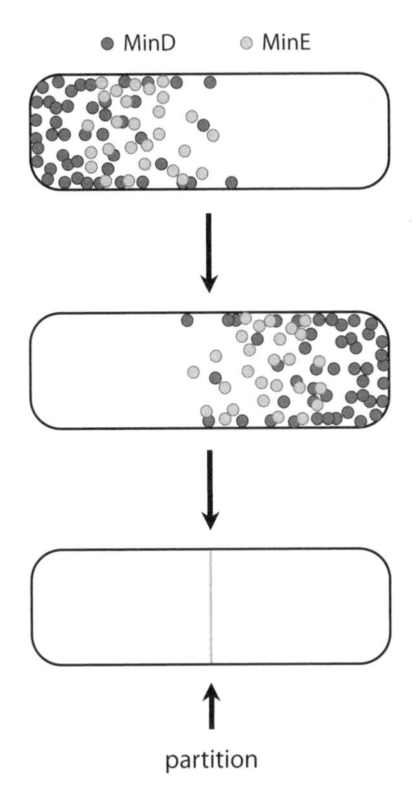

Figure 13. Policing the ends of an *E.coli* cell with Min proteins.

our familiar diagram of two feedback loops (fig. 14). The positive loop represents MinD binding to the membrane and promoting the binding of further MinD molecules. The negative loop represents MinD attracting MinE to the membrane, which then makes MinD fall off. We have all the ingredients needed for a pattern to emerge. Computer simulations show that with the appropriate molecular affinities and diffusion rates, this system can generate the observed oscillating pattern of Min proteins, with high concentrations continually moving back and forth from one end of the cell to the other every few minutes.

This example shows how our familiar double feedback loop can lead to a pattern within the confines of a single cell, creating distinctions between one region and another. For multicellular organisms we have a further level of patterning, this time among different cells of the same

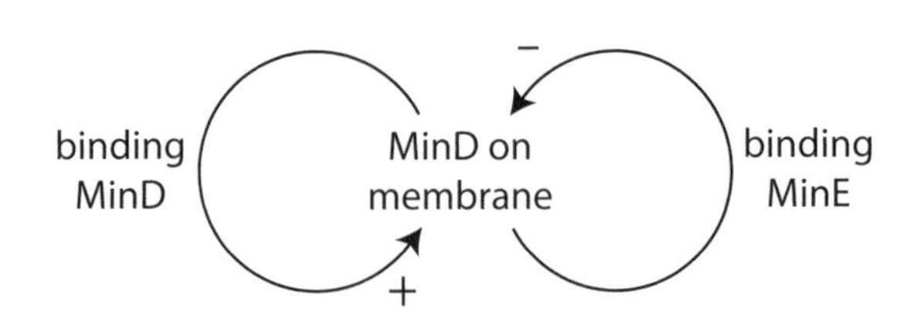

Figure 14. Interplay between reinforcement (positive loop) and competition (negative loop) for partitioning of the *E. coli* cell.

individual. As a fertilized egg divides and develops, various types of cells are generated, such as leaf and root cells of a plant, or muscle and nerve cells (neurons) of an animal. Within each broad category of cells there are further subtypes, such as the cells that form hairs or pores (guard cells) on a leaf, or the many different types of neurons in the nervous system (fig. 15). Many of these cells are polarized, with one end distinct from the other—the tip of a leaf hair cell is distinct from its round base, and each neuron has one end that receives inputs (dendrites) and another that sends an output (axon). As well as this internal pattern, the arrangement of cell types in the tissues and body is also highly organized, with cells distributed in particular spatial relationships. It is this process of patterning—the development of an egg into a highly defined and complex arrangement of cell types—that Turing wanted to explain. We have to look at more elaborate events to understand how this works, involving entire genes being switched on or off. So, before going any further, I first have to provide some extra background on how genes work.

Switching Genes On and Off

Recall that a gene is a stretch of DNA. In many cases the stretch can be roughly divided into two parts: a regulatory region and a coding region (fig. 16). Let's first look at the coding portion. This region contains information needed to specify the sequence of amino acids in a protein. However, the information is not read directly off the DNA. The DNA information is first copied, or transcribed, to produce another type of molecule, called RNA. RNA has a similar structure to DNA but is single rather than double stranded. If you think of the DNA in the nucleus of a cell as being like

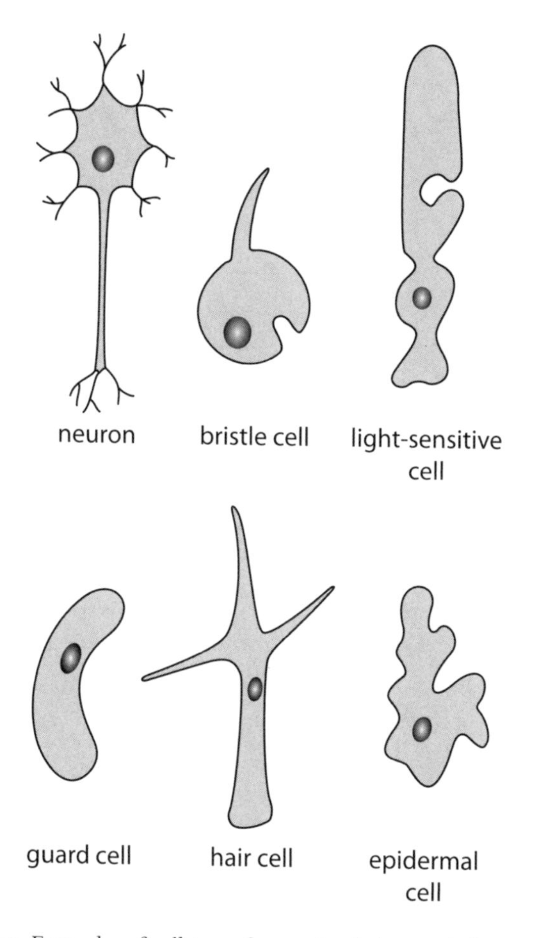

Figure 15. Examples of cell types from animals (*top*) and plants (*bottom*).

a set of books, with each gene being a page, then RNA corresponds to a photocopy of part of a page (the coding region). Many RNA molecules can be produced from the same gene, or many photocopies from the same page. After copying, the RNA molecules move out from the nucleus into the surrounding cell fluid or cytoplasm; here the RNA information is translated into the amino acid sequence of a protein. The protein then folds into a particular shape, depending on its sequence of amino acids. The whole process has been summarized with the saying, "DNA makes RNA makes protein."

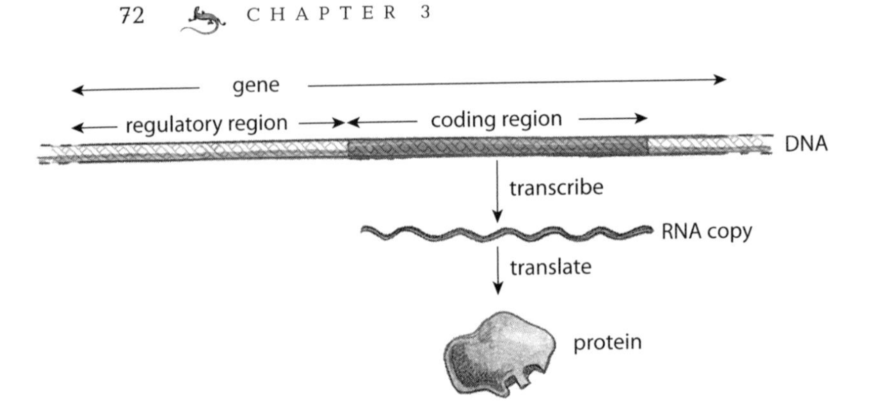

Figure 16. Typical gene showing how information in DNA is transcribed to make RNA, which in turn is translated to make a protein.

Although all cells of an individual contain the same DNA information, the genes being copied to make RNA may vary from one cell to another. It is like having a library with many rooms (cells), each of which contains the same set of books. Certain pages from the books are being copied in one room while other pages are being copied in another room. A gene being copied is said to be switched on, whereas if no copies are being made, the gene is said to be switched off. However, it is not a switch with only two positions—the rate of RNA copying can be set at different levels. What determines the extent of gene activity, the degree to which a gene is on or off?

This question brings us to the regulatory region. Figure 17 shows a simplified diagram of the regulatory region of a gene. You can see that it has several proteins, symbolized by apple and lemon shapes, bound to it. These are known as regulatory, or master, proteins. Regulatory proteins can recognize short stretches of DNA and bind to them. The apple regulatory protein binds to the DNA sequence symbolized by dark gray stretches, while the lemon regulatory protein binds to the light gray sequences. Once they are bound to the DNA, regulatory proteins can activate or inhibit the transcription of the gene, turning the activity of the gene up or down. They do this by influencing the RNA copying machinery which locks onto and transcribes the gene. Imagine that the apple regulatory protein promotes transcription of the gene when it is bound, switching the gene on. The apple protein would then be an activator of

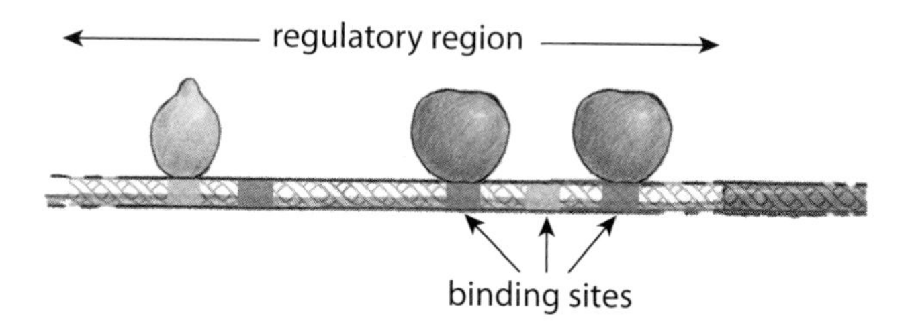

Figure 17. Proteins bound to regulatory region of a gene.

the gene. By contrast, the lemon regulatory protein may be an inhibitor, repressing the gene when it is bound, and tending to switch it off. The extent to which a gene is on or off depends on the balance between the activator and inhibitor proteins bound to its regulatory region.

You might be wondering where the regulatory proteins, the lemons and apples, come from. They are themselves coded for by other genes—the lemon protein by one gene and the apple protein by another gene. So, there are genes that code for proteins which in turn regulate other genes. As we will see, one consequence of these interactions between various genes and proteins is the generation of patterns.

A Molecular Fight

Many of the best understood cases of patterning are found in the fruit fly *Drosophila melanogaster*. This species is convenient for studying mutations and genes because it reproduces quickly and is easy to maintain in large numbers in the laboratory. Let's look at the formation of a particular type of cell, called a neuroblast, in a developing fruit fly. Neuroblasts are the cells that eventually give rise to the fly's nervous system. The upper panel of figure 18 shows a small region of the developing fly embryo where two neuroblasts are forming. You can see that there are initially two groups of light gray cells, and that a single neuroblast cell (dark gray) emerges from each. This provides us with a basic patterning problem: How does a cell become singled out from its neighbors to become a neuroblast?

A clue comes from mutant embryos in which the patterning goes wrong. The lower panel of figure 18 illustrates what takes place in a mutant which has a defect in a gene called *Delta*. Instead of a single neuroblast cell

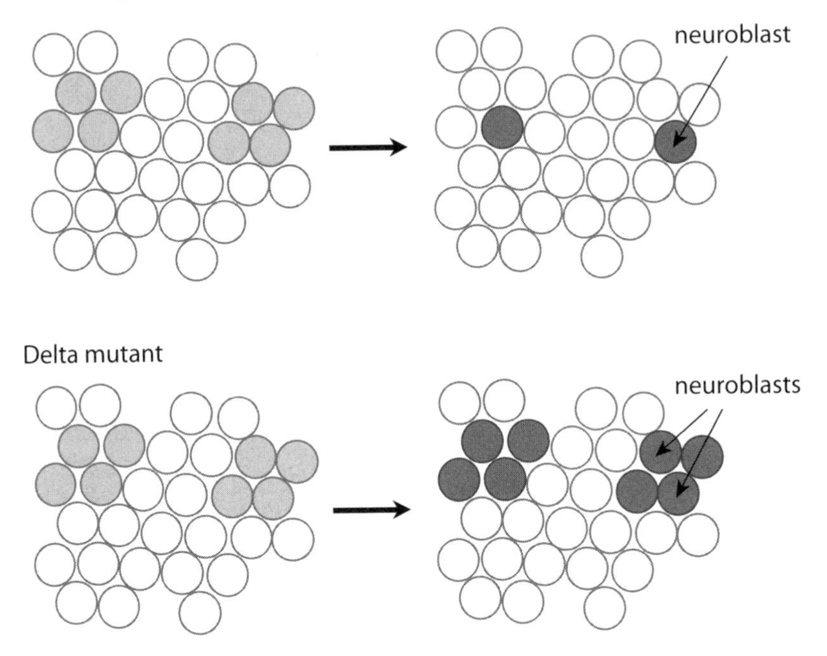

Figure 18. Neural patterning in fruit flies.

forming, all of the light gray cells turn into neuroblasts. Think of the light gray cells as normally fighting it out to see which one will become a neuroblast. Without the *Delta* gene there is no fight, so all of the light gray cells becoming neuroblasts.

To see how this works, look at the left part of figure 19 where two light gray cells are interacting with each other. You can see some arrows with positive and negative signs. The overall effect of these interactions is similar to our familiar positive and negative feedback loops. In each cell, the *Delta* gene (located in the nucleus) is switched on and produces Delta protein (by convention, protein names are not italicized, to distinguish them from the genes that code for them). The Delta protein molecules, symbolized by lollipop shapes, go to the cell membrane where they attach, sticking their heads out of the cell. As they jiggle about on the membrane they may collide with another protein, a receptor, located on the neighboring cell. The receptor has a shape which matches that of the Delta

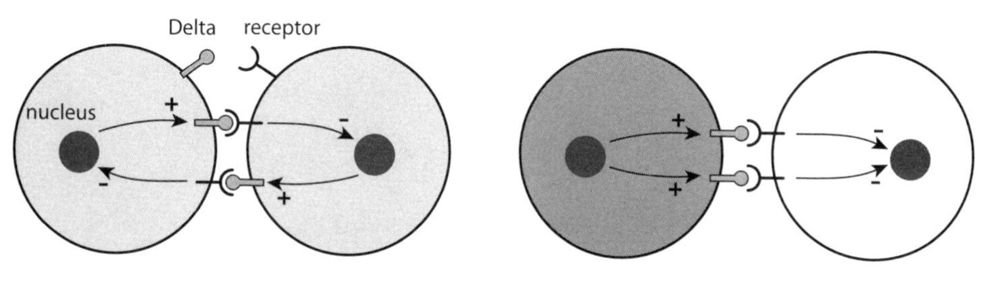

Figure 19. Signaling between cells involving the Delta protein and its receptor.

protein. If Delta binds to the receptor, it sets off a chain of events in the receptor-bearing cell, which involve molecules diffusing to the cell nucleus and switching off the *Delta* gene (negative sign in the figure).

Essentially, the production of Delta in one cell strikes a blow to its neighbor by trying to switch off its *Delta* gene. If the protein quantities are reasonably balanced, both cells switch off each other's *Delta* gene to the same extent. But if one cell has slightly more Delta, then this protein level increases further (it is reinforced) as it switches off the competition from its neighboring cell more effectively. Eventually one cell supersedes, and then contains lots of Delta while the other contains very little (fig. 19, *right*). I have illustrated the interaction between two cells, but if the same rules are extended to a small group of cells, a similar result can be obtained—eventually one of the cells contains far more Delta protein than the others. The principles involved are similar to those that Turing invoked, but instead of simple chemical reactions in a solution, we are dealing with chains of events in which cells are signaling to each other and genes are being switched on or off.

Another feature emerges from the Delta story. Notice that the white cell on the right of figure 19 has acquired an asymmetry, or polarity—it only has Delta signals coming in at its left side. Not only can such patterning events lead to distinctions between cells, they can also provide a way of orienting the internal organization of cells. As we have seen, many cell types, such as the hair cells on a leaf or the bristle cells on a fly wing, are highly polarized and exhibit differences from one end of the cell to the other (fig. 15). Moreover, these polarities are often organized in relation to other cells—a leaf hair cell points outwards, away from the internal cells of the leaf, and the bristles of the fly all point in a defined direction along the animal's body. Such polarities also trace back to signaling among cells, involving similar processes to those we have already encountered.

Looking into Gradients

We have seen how interconnected feedback loops can lead to the transformation of patterns, but I want to now turn to another type of patterning mechanism that at first sight seems to be based on different principles. The upper part of figure 20 shows a line of cells in which there is a gradient in the concentration of a molecule, with a high concentration at the left end (black) and a gradual decline to a low concentration at the right (white). You could imagine such a gradient might arise by molecules being produced by the cells at the left end and diffusing to the other end. Now suppose that these molecules can influence the activity of a gene, such that the gene is switched on if the concentration of the molecule is above a threshold level. We would end up with two distinct regions: a region to the left where the gene is on (the concentration of our molecule is above the threshold), and a region to the right where the gene is off (fig. 20, *lower panel*). Instead of a gradient we have a steplike pattern in gene activity: high in the left half and low in the right half. We have gone from a relatively smooth initial pattern to a more sharply defined one. It seems as though we do not have to invoke the principles of reinforcement or competition here. All we need to generate a steplike pattern is diffusion creating the gradient, and a threshold level affecting gene activity. But how are thresholds actually measured? It is easy for us to draw a line and say that whatever is above the line behaves differently from what is below, but how do cells do the equivalent? To answer this question, I want to look at one of the best studied examples of gradients.

The story that follows involves the interaction between two genes, called *bicoid* and *Hunchback*. For the sake of simplicity, the proteins produced by these genes are symbolized by two fruits—apples for Bicoid and pears for Hunchback. Let's start with the apples. During its life, a female fruit fly produces numerous egg cells, each about half a millimeter long. After fertilization, the egg nucleus undergoes several rounds of division. The egg cell then contains many nuclei. It is rather unusual for nuclei to divide like this within a cell but it is only temporary; later, the nuclei are packaged into separate cells. As the nuclei multiply, a graded

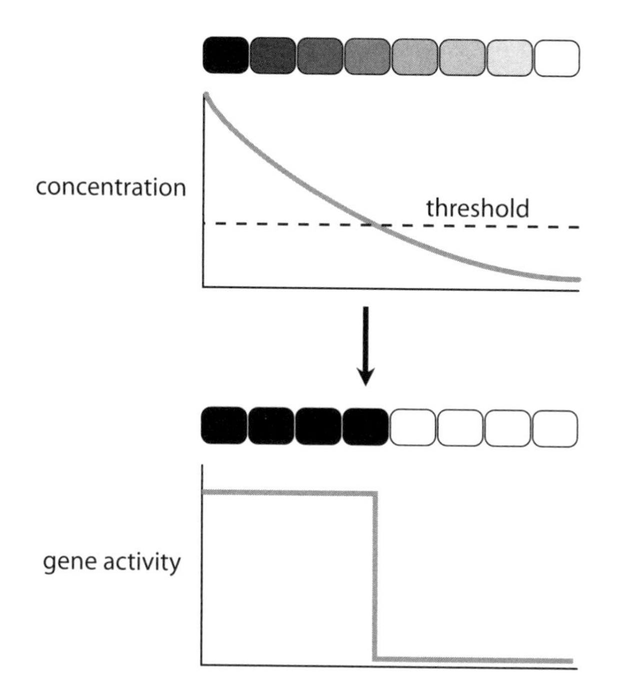

Figure 20. Gradient and threshold leading to a steplike pattern of gene activity.

distribution of the apple protein (Bicoid) is established in the embryo, with a high concentration at the head end and a low concentration at the tail end (fig. 21, for the time being, we will ignore the issue of how this gradient arises). In some nuclei, the apple protein then switches on our second gene, *Hunchback,* and this leads to the production of pears. In other words, the apples bind to the regulatory region of the *Hunchback* gene, switching it on. This chiefly happens in the nuclei at the head end where the apples are most abundant. The result is that the pears at this end are also abundant (fig. 22). But notice that in switching on pear production, the gradient of apples has created a more steplike pattern of pears. The curve for pear concentration is flat near the head end and then drops more steeply than the gradient in apples (*lower panel*, fig. 22). Here, then, is an example of a graded distribution of one protein that results in some sharper distinctions for another protein during development.

To understand how this steep drop-off is generated, we have to look more closely at the regulatory region of the *Hunchback* gene. This region

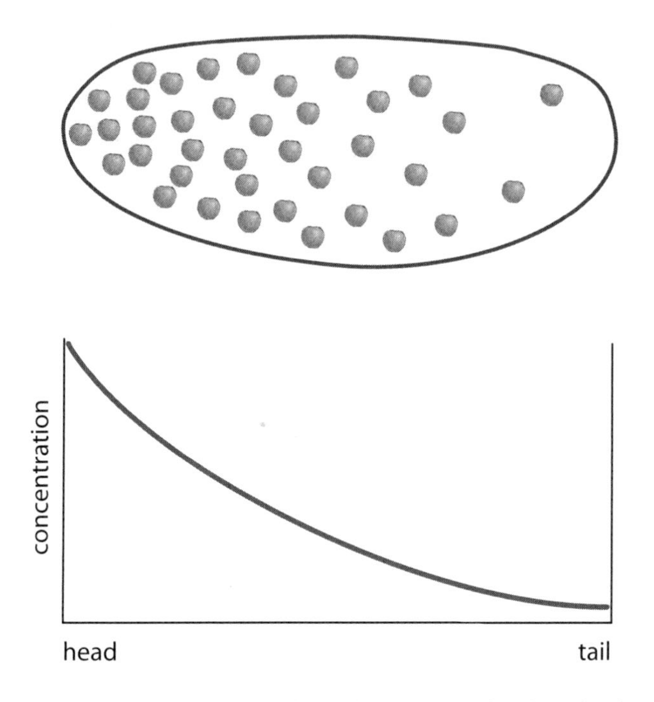

Figure 21. Gradient in concentration of Bicoid protein (apples) from head to tail in a fruit fly embryo.

has several sites to which the apples (produced by *bicoid*) can bind (fig. 23). The key lies in the way these apples interact with each other. When an apple binds to a site, it promotes the binding of another apple to one of the nearby sites—the apple proteins boost each other's attachment. To see how this generates the pattern of pears, let's start at the tail end of the embryo. Here there are very few apples, so few of the sites in the *Hunchback* regulatory region have apples bound to them. The gene is therefore mainly inactive and produces very few pears. But as we move toward the other end, the concentration of apples rises until we reach a point where there are enough for several to start binding. And because they promote each other's attachment, the mutual assistance between apples results in a rapid increase in pear production. This is one of the main reasons that you see a steep rise in the concentration of pears in the middle of the embryo. Further toward the head end you might expect the level

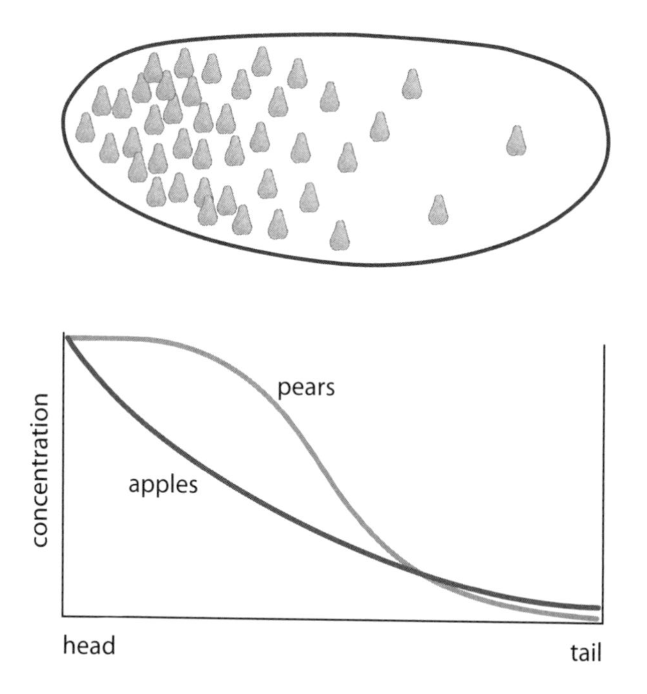

Figure 22. Steplike pattern in concentration of Hunchback protein (pears) from head to tail in a fruit fly embryo.

of pears to grow higher and higher. But there is a limit to the number of sites to which apples can bind, so the apples start competing for the few remaining positions. Eventually, all the apple sites are occupied, and we then reach an upper limit for pear production. This explains why the pear curve flattens out toward the head end.

We are back to the same ingredients for pattern transformation that we have seen before. We have a population of apple proteins diffusing in the embryo and binding to their sites. There is reinforcement as the apples assist each other in binding to multiple sites in the regulatory region. Competition arises because of the limitation in the number of sites to which apple proteins may bind. The patterns can persist because of the stability of protein binding, and diffusion is not so great as to dissipate the arrangement of pears and apples. The overall result is a transformation of one pattern into another: an apple gradient into a sharper curve of pears.

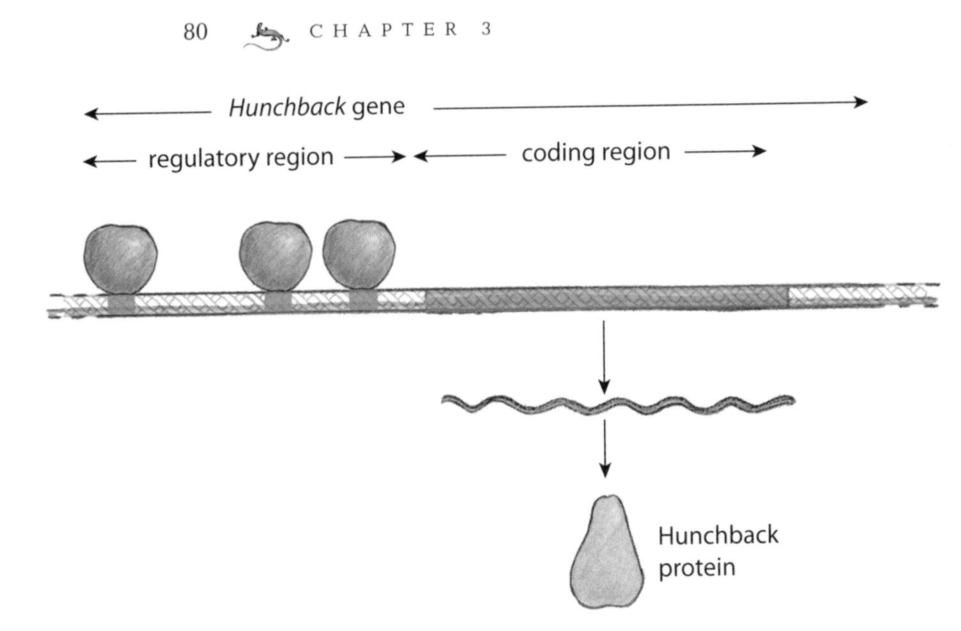

Figure 23. Bicoid proteins (apples) bind to regulatory region of the *Hunchback* gene, increasing its activity, and increasing the level of the Hunchback protein (pear).

A Common Form

All the examples of patterning we have encountered in this chapter have some common, basic features. Each involves interconnected loops of reinforcement and competition, fueled by a balance of population variation and persistence: the clustering of conversations at the tea party, Turing's reaction-diffusion system, the partitioning of an *E.coli* cell, the molecular fight for Delta supremacy in neuroblasts, and building on the Bicoid gradient of the early fly embryo. All of these are based on this same basic recipe. The overall result is that patterns are refined; activity becomes more clustered or a distribution is sharpened. If we were to look at other examples of patterning in development, we would continually encounter the same basic ingredients. The particular genes and molecules may vary in each case, but the principles would be the same. Why is it that we always find this same arrangement?

To refine a molecular pattern, some components must increase while others decrease in a coordinated manner. Reinforcement alone would simply result in components boosting themselves to ever higher levels.

Conversely, competition alone would result in general consumption of components (substrates), or inhibition that lowers them everywhere. However, the situation is much more interesting when reinforcement and competition are linked together. Instead of a general increase or decrease, some components can rise at the expense of others. If we also incorporate the effects of distance through molecular signaling or diffusion, then the components may become organized in a spatial manner, and occupy distinct regions. Components may increase in abundance at some locations, but not at others. In this way, a pattern can be refined, sharpening a boundary or activating a gene in a subset of cells.

The interplay of population variation, persistence, reinforcement, and competition can be compared to what we encountered with natural selection in chapter 1. For development, population variation involves statistical interactions among numerous molecules or cells within an individual. In the case of evolution, population variation involves mutations continually cropping up and being shuffled in a population of individuals. For both processes, population variation is necessary for change to take place.

Too much variation, however, is a problem. The degree of variation has to be constrained; otherwise any changes would dissipate as soon as they took shape. Persistence is established during development through chemical stability and limits on molecular mixing. Cell membranes, for example, prevent the contents of one cell from mixing too readily with those of its neighbors. The membranes also help the cell contents, including regulatory proteins, stay together and get passed on to daughter cells, allowing patterns of gene activity to be maintained. Similarly, the diffusion rates of molecules that signal within and among cells are constrained within particular bounds. In evolution, persistence involves the faithful transmission of information from one generation to the next through reproduction and gene replication. In both development and evolution, persistence is needed as a balance to variation. Without variation nothing would change, and without persistence, anything that did change would not last.

Population variation and persistence are essential ingredients, but left to their own devices they do not lead to interesting outcomes. For organization to emerge, some components must increase at the expense of others. This takes place through interplay between reinforcement and competition. In development, reinforcement occurs as molecules boost their own production or enhance each other's action. The increase in these molecules is counteracted by their switching on inhibitors or by competition for limiting factors. This relationship between reinforcement and competition provides the driving force for patterning. In evolution, reinforcement involves genes promoting their own replication by influencing survival and reproduction. Reproductive success then leads to an increase in competition because of limited resources. In both cases, it is the coupling between reinforcement and competition that leads to organized change.

Notwithstanding the many differences between development and evolution, we see the same principles and interactions. You might not be too surprised that the principles I use to describe evolution are also be found in my account of development. As I mentioned at the start of chapter 1, this particular collection of principles was arrived at by looking for features that were common to several living transformations, not just one. What is perhaps more surprising is that the commonalities we find are not superficial, but are at the heart of each process. They define the core interactions that lead to transformations in each case. This suggests that evolution and development do not operate in completely distinct ways but are different manifestations of a common set of underlying principles through which organization can emerge.

We should be careful, though, not to take such comparisons too far. For example, while mutations play a key part in evolution, there is no exact counterpart to mutation in development. The process of mutation involves a one-off heritable change in a large DNA molecule; this is very different from the collisions of molecules involved in reactions and diffusion. Also, reinforcement in evolution involves replication, while molecules can reinforce their actions during development without being directly copied. Comparisons can be confusing rather than helpful if taken too literally, as with the example I gave in the introduction of a teacher comparing a horse with a knight in chess. War and chess exhibit similarities in form rather than a detailed correspondence. Similarly, we should not try to match the components of evolution and development too precisely. We are dealing with a similarity of form, with similar principles and feedback loops, not detail. Development and evolution involve the same basic recipe but are dressed in two very different guises.

In conveying the basic principles of development, I have, however, glossed over many issues. Each of the transformations I have described assumes a prior context: a region of cells that form neuroblasts, or a gradient of Bicoid protein in the egg. But how are these contexts themselves generated? And how do a few pattern transformations relate to the development of complex multicellular organisms like apple trees and humans? To answer these questions we must look further into how development takes place.

FOUR

Completing the Picture

SOME OF CÉZANNE'S PAINTINGS seem unfinished. In *The Garden at Les Lauves* (fig. 24, plate 4), Cézanne stopped after sketching in a few areas with color. Why he left the painting this way is unclear—perhaps he reached a point where he could go no further, or maybe he was just happy to leave it as it was. Whatever the reason, the painting reveals how he built up texture and pattern. He seems to have started with a preliminary color sketch that he then built upon by applying further patches of color. Each stroke of color depended on the colors already on the canvas, and in turn influenced the next colors he applied. Cézanne's description of portrait painting also suggests this highly interactive process: "If I weave around your expression the infinite network of little blues and browns that are there, which marry together, I will make you look out from your canvas as you do in life.... One stroke after another, one after another."

In the previous chapter we saw how patterns can be transformed during development. An initial distribution can be modified to give new patterns of gene activity, such as a spotty or a steplike arrangement. This is rather like the way an artist might add a few extra strokes of color to a canvas, transforming its appearance. But a few strokes by themselves are not enough to create a complex pattern. The picture has to be built up by placing color upon color, stroke upon stroke. The development of an embryo does not involve just one or two transformations but instead involves the accumulation of many changes, each building on the other. How does this process of continual elaboration work?

Figure 24. *The Garden at Les Lauves*, Paul Cézanne, 1906. See plate 4.

An Embryonic Cocktail Party

Think of a developing embryo as a canvas upon which colors and patterns are placed. As we saw in the previous chapter, the early fruit fly embryo has a gradient of regulatory proteins from head to tail, symbolized by apples. This is equivalent to having an early wash of color on our canvas, say apple green, that gradually fades from one end to the other (fig. 25, *top*). This gradient of apples is then built upon to generate a steplike pattern of pears. This happens by the apple proteins binding to the regulatory region of the pear gene, and switching it on to different degrees along the embryo (fig. 25, *middle*). We now have two colors on our canvas,

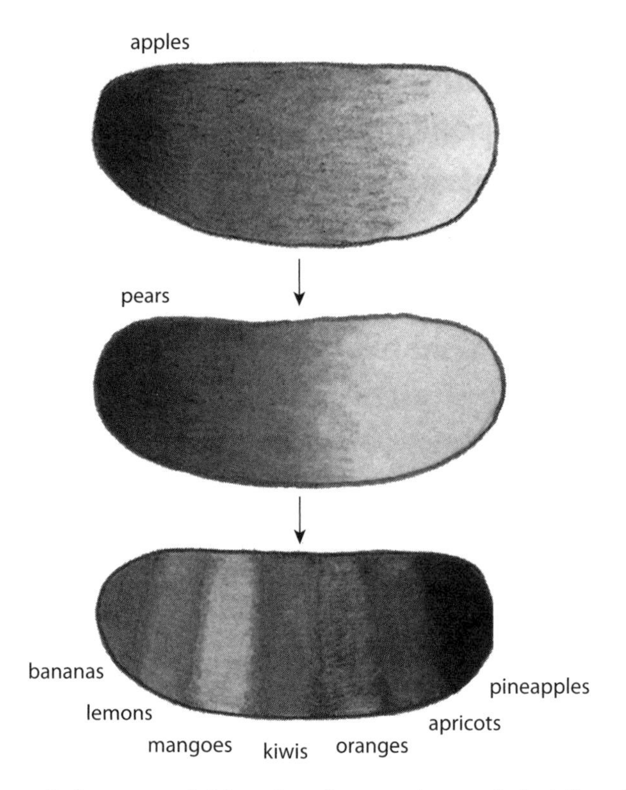

Figure 25. Refinement and elaboration of patterns in an early fruit fly embryo.

a shallow gradient of apple green and a more steplike pattern of pear green. As in painting, the second color has been placed in relation to the first.

We can continue, adding further colors to our canvas. Apples and pears are regulatory proteins with the ability to switch genes on or off. Once our apple-green and pear-green washes are on the canvas, further genes can be activated by them. The products of these genes can be symbolized with additional fruits, such as bananas, lemons, mangoes, kiwis, oranges, apricots, and pineapples. That is, we have a gene that makes banana proteins, another that makes lemon proteins, and so on (these genes correspond to what are known as gap genes).

What follows is an elaborate interplay among the various fruit genes, similar to the interaction of colors on a canvas. Just as each stroke applied by an artist both depends on and influences other strokes, each gene both contributes to and responds to other genes. This takes place because each gene not only codes for a fruit (regulatory protein), but also has a regulatory region to which fruits can bind. The result is that some genes and their regulatory proteins tend to reinforce each other's activity while others compete. New patterns start to emerge through a territorial battle. Bananas are restricted to one end, followed by a band of lemons, then mangoes, and so on, to pineapples at the other end (fig. 25). The principles governing this process are no different from those we encountered in the previous chapter. Through molecular jiggling, persistence, reinforcement, and competition, patterns are gradually transformed. Now, however, many interactions are taking place in parallel. The overall result is that after starting off with only a gradient of apple green, the pattern on our canvas has been built upon to produce several different bands of color. But no external artist is driving this process; instead, it has emerged through molecular interactions.

A Cooperative Effort

Underlying these events in the embryo is the way various regulatory proteins (signified here by fruits) interact when they bind to regulatory regions of genes. Each regulatory region may contain ten or so different sites to which particular regulatory proteins may bind. The regulatory region of one gene, for example, might have sites for apples, pears, bananas, lemons, and mangoes. Such a regulatory region acts as a statistical magnet that tends to bring these fruits together. Having been brought into proximity, they can easily jostle and interact with one another, and these interactions determine whether the gene gets switched on or off. By bringing fruits close together, regulatory regions encourage interactions. Consequently, you can not predict what any one regulatory protein will do without considering how it operates in relation to other regulatory proteins. For development to proceed in an organized fashion, the overall result of these interactions is the coordinated activation or inhibition of genes as the embryo develops. This is the principle of cooperation operating at the level of gene regulatory regions.

Cooperation does not only operate at the level of regulatory regions. Recall that as the fruit fly embryo develops, the nuclei within are dividing. The fertilized egg starts with only one nucleus, but by the time our bands of bananas, lemons, and other fruits have formed, many nuclei are in the embryo, each containing its own copy of the DNA. Having multiple nuclei in close proximity allows them to interact. Regulatory proteins produced by one nucleus, say lemons, can move to nearby nuclei and influence the regulatory proteins they produce. This conversation between nearby nuclei is another form of cooperation that allows development to proceed in an organized manner.

Having multiple nuclei, rather than a single nucleus, within one cell is a peculiarity of the early fruit fly embryo. In most animals and plants, cells of the early embryo divide, and their nuclei also divide at the same time to produce multiple cells, each with an outer membrane and a single nucleus. These multicellular embryos are patterned by the same sort of principles as for the fly, except that additional molecules are needed to send and receive signals across the membranes (operating in a similar way to the Delta signaling and receptor system described in the previous chapter (on pages 73–75). Again we have local cooperation, this time with nearby cells in the embryo interacting with each other. Interactions of this type among cells also take place in the fruit fly at later stages, when the multiple nuclei of its embryo become surrounded by membranes, and form separate cells. Development is always a cooperative effort, involving interactions between nearby cells, nuclei, and regulatory molecules.

Regulatory Riches

When several different components interact, many outcomes become possible. The vast array of molecules in the universe comprise different combinations of only a hundred or so atoms. And all genomes on earth are different combinations of just four bases A, G, C, and T. This principle of combinatorial richness also applies to gene regulation. So far, we have encountered nine regulatory proteins, symbolized by apples, pears, bananas, and so forth. But this is just the tip of the iceberg. Each plant or animal genome may code for hundreds, or even thousands, of regulatory proteins. The human genome, for example, contains around 25,000 genes. Of these, about 2,500 are thought to code for regulatory proteins that can switch other genes on or off. That is, the human genome codes for about 2,500 different types of fruit.

To see what this can lead to, let's assume that a genome codes for one thousand different types of regulatory protein. A given cell, or nucleus with its surrounding cytoplasm, produces only a fraction of these proteins at any one time, say 10 percent. This means that each cell produces one hundred out of the thousand possible regulatory proteins, a selection of one hundred fruits from a range of one thousand. Another cell might have a different selection of regulatory proteins, a different combination of one hundred fruits. How many different types of cell, in terms of regulatory protein production, can we potentially have with this system? The answer is around 10^{300} (a one followed by three hundred zeros), which is a colossal number. We are dealing with a vast hyperspace of 100 dimensions (the selection of fruits), each with 1000 possibilities. This is the principle of combinatorial richness applied to regulatory proteins.

But combinatorial richness doesn't stop there. If we have many different types of cells next to each other, yet more possibilities open up. We can arrange the cells in various ways, just as we may arrange chess pieces in different ways on a chess board. Unlike chess though, which has only twelve types of pieces (six black and six white), we have 10^{300} possibilities to play with. And instead of sixty-four places on the board we may have many thousands or millions of cells. The number of possible combinations does not even bear thinking about. We have a vast hyper-hyperspace of possible cell types and arrangements. I call this great space of possibilities *developmental space*.

Although developmental space is unimaginably large, we can nevertheless visualize it as a very large three-dimensional space. Just as we pictured evolution as populations traveling through a vast genetic space (chapter 2), we can think of development as an embryo voyaging through an enormous developmental space. At fertilization, the embryo comprises a single cell with a particular combination of regulatory proteins, occupying one position in developmental space. Selections of regulatory proteins are then activated in various regions of the embryo as the nuclei and cells divide, corresponding to the embryo taking a particular path through developmental space. The example of the fly embryo going through its series of fruity patterns represents the beginning of one such journey. The fly embryo is negotiating a narrow path through a vast space of regulatory possibilities, just as Cézanne painting a picture represents a journey through an enormous space of color possibilities. So far, we have looked at a few steps on this journey, or a few examples of how one pattern of gene activities can be transformed into a slightly different pattern. To get a better sense of what this journey entails, let's continue a little further with the story of the fruit fly.

Building on the Past

At the top of figure 26 you can see the fruit fly embryo discussed earlier, with banana proteins at the head end and pineapples at the tail end. During the next stages of development several things take place. One is that the nuclei of the embryo become surrounded by cell membranes, forming separate cells. Another is that a repeating pattern of gene activity begins to emerge, as shown on the right side of figure 26. Several new regulatory proteins are produced, which are symbolized with three fruits: strawberries, tangerines, and grapes. Where there was previously only one color, representing bananas, for example, there is now a thin strip of strawberries, followed by a strip of tangerines, a strip of grapes, then another strip of strawberries, tangerines, and grapes. The same strawberry-tangerine-grape theme is played again and again along the length of the embryo. These proteins are produced by a further set of genes (known as segment polarity genes). Their repeating pattern of activity is generated by building on the earlier distribution of bananas, lemons, and so forth, according to the principles we have already encountered. But in this case, the genes are switched on at repeated locations along the length of the embryo.

At the same time that this repeating theme is being established, a further set of regulatory proteins are being produced, as shown on the left side of figure 26. This pattern of proteins is not so different from our earlier pattern of bananas, lemons, etc. The main difference is that the boundaries between the regions have become sharper and better defined. These regulatory proteins are produced by yet another set of genes (called Hox genes).

Now imagine superimposing our two patterns, as shown in the bottom embryo of figure 26. The result is a theme with variations—a repeating pattern combined with modulations given by the progression of sharply defined domains from head to tail. This overlapping pattern provides a

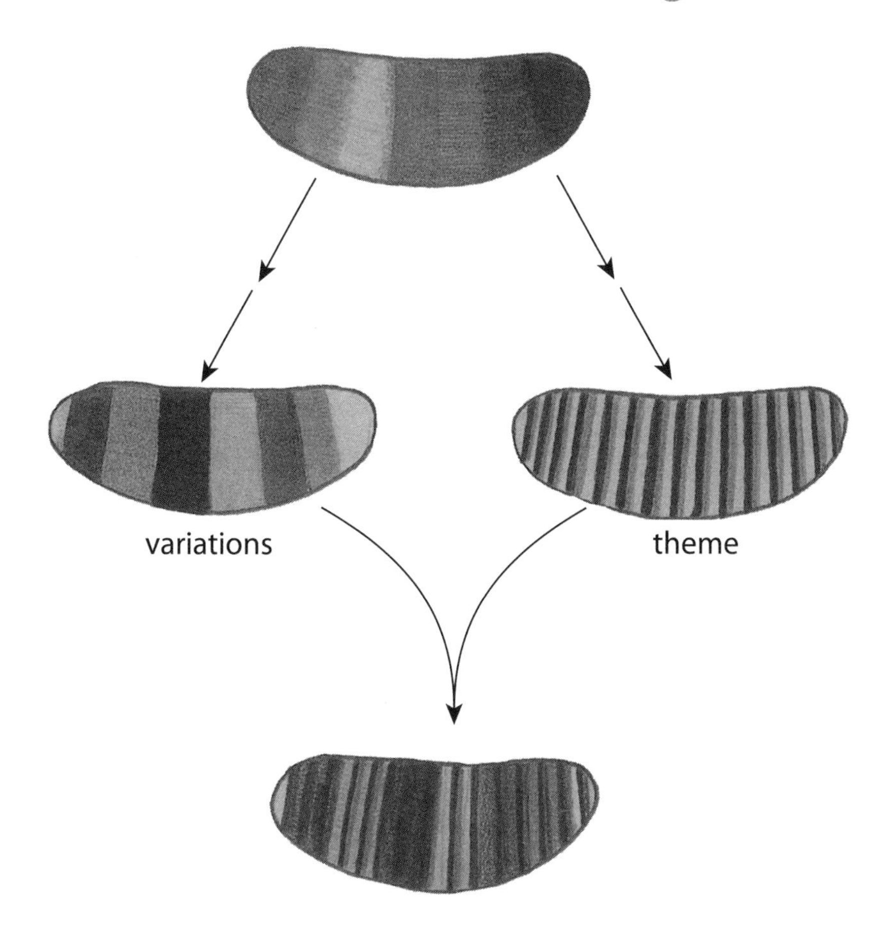

Figure 26. Interactions between themes and variations in a fruit fly embryo.

basic framework for the fly. Each adult fly has fourteen segments, some of which form the head, others the middle section (thorax), and the rest the tail section (abdomen). The segments vary from one to the next: one segment bears antennae, others bear legs, one bears wings, and so on. We have a repeating theme, the segment, and variations because the segments are distinctive. This organization traces back to the pattern of regulatory proteins in the early embryo, our overlapping pattern of fruits. This early combination of regulatory proteins defines the modular arrangement of the fly, setting up patterns of gene activity that lead to the segmented organization of the adult. Mutations that disrupt the genes coding for the regulatory proteins will result in flies developing with missing segments, or with one type of segment replaced by another.

All of these patterns arise through repeated application of the processes we have already covered. Each pattern of regulatory proteins is built upon to generate a further pattern, and this happens again and again, transforming the pattern. Development does not involve just one or two transformations; it incorporates a whole series of changes that recurrently elaborate the picture, always building on what went before.

What drives the embryo along this particular path through developmental space? The answer has to do with the ever changing context generated through patterning. Each patterning step not only depends on what went before, it also changes the context and thus influences the next steps. This takes place because regulatory genes can both respond to other proteins through their regulatory region, while also affecting the activity of other genes by producing their own regulatory proteins. And, because the regulatory proteins from one nucleus or cell may influence its neighbors through molecular signaling, this process is coordinated in space and time. Cells respond according to a particular context, while also modifying the context, driving the embryo forward on its developmental journey. In this way, the embryo forges its path through developmental space. It is the principle of recurrence, with the context being continually redefined through repeated application of the same underlying process.

If each stage in embryological development is a modification of its previous context, then you might reasonably ask where the first context comes from. In the case of the fly embryo, we began our story with a gradient of Bicoid protein, symbolized by a gradient of apples from head to tail. How does this gradient arise? The answer has to do with the mother fly. The egg does not form in isolation within the mother, but is connected at one end with cells that supply various components; these include RNA copies of the *bicoid* gene (recall that RNA is an intermediate molecule needed to make a protein). These RNA copies are initially anchored at the head end of the egg where they are used to make Bicoid protein (apples). The RNA and Bicoid protein then propagate through the egg, forming a gradient from head to tail. So the gradient of apples in the early embryo traces back to an earlier context: the egg and its surrounding cells in the mother.

As we trace the context of the embryo further back we find ourselves in the context of the mother.

But you may ask what patterns the egg and the nearby cells of the mother. This in turn depends on the way the mother fly herself developed. The starting point of an organism's development can also be seen as an outcome of development in the previous generation. Rather than having a clear beginning, our journey through developmental space forms a cycle, with the journey of each adult providing the context for the journey of its offspring. Like the story of *The Odyssey*, it is a journey that returns to its origins. This is sometimes called the chicken and egg problem. Which came first: the chicken or the egg? But this is only a problem if we want to define an ultimate starting point. It is rather like saying that a circle is a problem because we can't define a beginning or end. This is not really a problem; it is simply the nature of a circle. Similarly, the embryo's path through developmental space forms a loop, with one journey around the loop leading to the next. The key issue here is not to search for a precise beginning, but to understand how the embryo is driven along its path. As we have seen, this developmental path involves the recurrent application of the same set of underlying principles, with the embryo being continually driven on by its context while changing its context at the same time. Through this process, the embryo is transformed from simple beginnings to an elaborate, multicellular individual.

We have seen how recurrence can lead to the elaboration of patterns within an embryo. But how can we account for patterns at numerous and different scales, from the overall organization of the body to the detailed arrangement of cells in the eye or a leaf? To answer this question we have to look more closely at something that has been going on in the background of our developmental journey: growth.

The Expanding Canvas

George Stubbs was obsessed with horses. Stubbs was born in Liverpool in 1724, and initially trained as a portrait painter but his focus gradually shifted from people to horses. He developed a unique reputation for painting these animals and won commissions to depict them with their owners, or grouped together at a hunt or on the racing track. It is therefore no surprise that when an exotic striped horse first arrived in England from the Cape of Good Hope in 1762, Stubbs was soon on the case (fig. 27). Stubbs' portrait is the most accurate and detailed rendition of a zebra for its time, even though the animal looks slightly out of place in an English woodland.

In producing such an accurate depiction, Stubbs probably used a variety of paint brushes. Perhaps he started by using relatively large brushes to define the overall arrangement of the picture and then used increasingly fine brushes to introduce greater detail. By varying the brush sizes in this way, he could capture different scales of organization, from broad composition to fine distinction. How does this process of refining brushwork compare to what happens during the development of a real zebra?

There are several species of zebra with different numbers of stripes. The common zebra (*Equus burchelli*) has about twenty-five stripes, the mountain zebra (*E. zebra*) has about forty stripes, and Grevy's zebra (*E. grevyi*) has about eighty stripes (fig. 28). (Judging by the number and pattern of stripes on the animal, Stubbs painted a mountain zebra.) Jonathan Bard has proposed that the variation between the different species of zebra is related to the time at which patterning happens in the embryo. The patterns underlying the stripes are thought to first develop in zebra embryos when they are a few weeks old, as shown on the left side of figure 28. Bard proposed that in all species, these initial stripes are spaced at regular intervals, with about 0.4 mm from the middle of one stripe to the middle of the next. If the initial stripes arise at twenty-one days, when the embryo is relatively small, only about twenty-five stripes spaced at 0.4 mm intervals could fit along its length; this is the case in the common zebra. But if the pattern is established later, when the embryo is larger, more stripes could be included, resulting in the patterns seen on the Mountain or Grevy's zebra.

This example highlights interplay between developmental patterning and growth. As development proceeds, the number of cells rapidly increases and this is often associated with an increase in size. It would be as if a canvas is continually expanding while an artist is working on it. Although this may sound more complicated than painting on a fixed canvas, in one way it makes things simpler. Instead of needing a series of increasingly fine brushes to insert greater detail, it is possible to use the same size brush all the time; the same brush that provides broad distinctions

Figure 27. *Zebra*, George Stubbs, 1763.

when the canvas is small allows much finer distinctions to be depicted when the canvas is large. This is exactly what happens in the example from zebra development. At the time the stripes first appear, they are always spaced about 0.4 mm apart. What varies is the size of the embryo, so that if the stripes appear later more can fit along its length, resulting in finer subdivisions on the adult animal.

The same interplay between growth and patterning applies to many other aspects of development. Although the pattern of regulatory proteins becomes finer as the fly develops, the number of nuclei or cells is continually increasing, so the scale of patterning in terms of cells does not change much. This means that similar principles of local interaction and signaling among cells can operate recurrently and give rise to patterns at

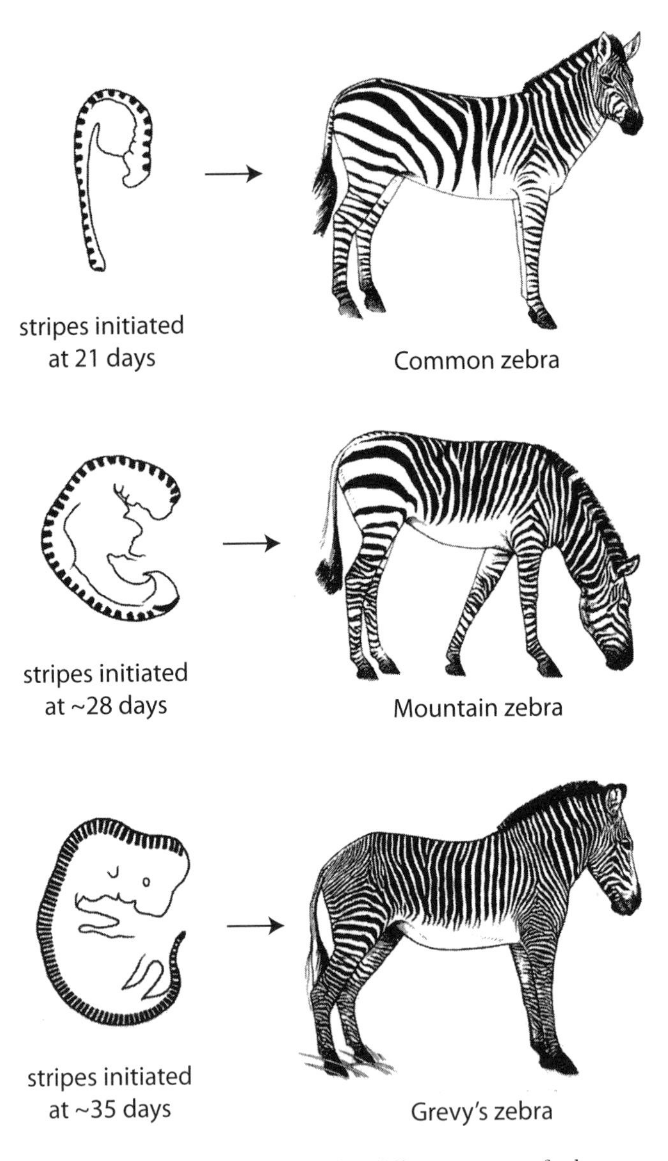

Figure 28. Stripes generated in different species of zebra

many scales as development proceeds, just as a single brush size can be repeatedly used to depict various levels of detail on a growing canvas.

To better understand this interplay between growth and patterning, imagine you are in the nucleus of a developing embryo, trying to decide whether to switch particular genes on or off. This decision is always relational, based on local conversations with your neighboring cells and nuclei. At the early stages of development when only a few cells are present, your neighbors constitute a large part of the embryo, and you effectively acquire information about what is happening in the embryo as a whole. Based on this information you might decide that you are at the head end, and then switch on the appropriate genes. But as development proceeds and more cells arise, your neighbors occupy a smaller and smaller fraction of the embryo and your sense of the whole picture is increasingly restricted. Consequently, you make decisions at a finer level of detail. Having previously decided that you are at the end of the embryo where the head will form, you now try to decide where you are in the head. Perhaps you work out that you are near the front of the head, where the eyes form, so you switch on the next set of genes, appropriate to this location. At each stage you commit yourself to a decision and subsequently build on this through local interaction. Because cells continually grow in number while embryos are being patterned, no new mechanisms need be invoked to account for the increasing level of detail that emerges.

Deformation

We have now covered many of the key principles of development, but there is another issue we must tackle in order to complete the picture. There is more to development than patterning and a change in scale. The adult is not just a magnified version of the egg; otherwise we would simply be large spheres bouncing around. Instead, plants and animals grow or contract preferentially in certain places and in certain directions as they develop. For example, a leaf emerges from the side of a stem or a leg grows out from a body. Our canvas does not grow uniformly, but expands or contracts in particular ways.

To get a sense of what this means, look at figure 29 (plate 5). The top shows a photograph of Cézanne's *Apples and Biscuits*, which depicts some red (darker gray in the black and white photograph) and yellow (pale) apples on a table. Imagine that this canvas is made of a special material that grows at a particular rate. Over time the canvas grows bigger and bigger according to the rules of compound interest. Certain colors on the canvas, such as yellow, also have the magical property that they increase the local rate of growth. The lower part of figure 29 illustrates the result when the growth of such a canvas is simulated in the computer. You can

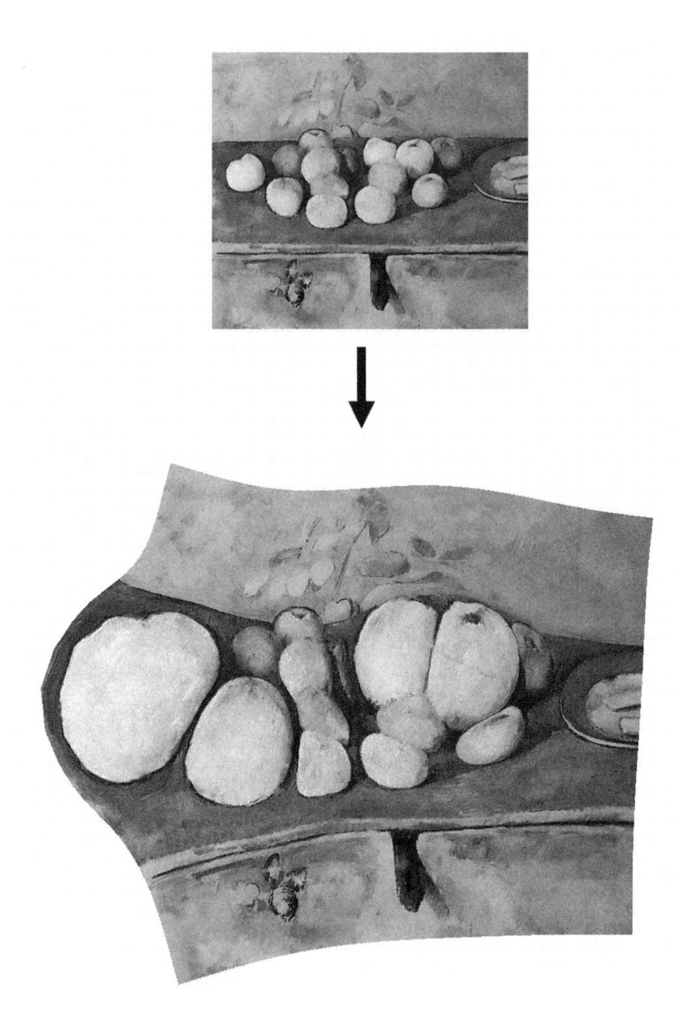

Figure 29. *Apples and Biscuits*, Paul Cézanne, c. 1880, and a deformed version caused by increased growth of yellow regions. See plate 5.

see that the overall size of the canvas has increased and that it has also become deformed by the disproportionate growth of the yellow (pale) apples.

Notice, though, that the yellow apple on the left grew larger than the other yellow apples. Why is this? The answer has to do with reinforcement and competition. Yellow regions on the canvas tried to grow more than other regions. Left to their own devices, the yellow apples would grow larger and larger, continually reinforcing their advantage in size through the law of compound interest. But the yellow apples are not by themselves; they are connected to the rest of the canvas. Consequently, if a yellow apple gets larger and is surrounded by areas that are not yellow, it tries to invade and stretch its neighboring regions, forcing them to grow larger to accommodate its increasing size. However, neighboring regions resist this stretching as they try to grow at their own pace. This conflict between the yellow apple and its surroundings creates forces or stresses in the canvas that counteract the yellow apple's growth. There is essentially a competition between the yellow apple and its neighbors for space. The result is a compromise in which the yellow apples grow less than they would like to, while nearby regions expand more than they would otherwise. The yellow apple on the left is less constrained by slow-growing material because it lies near the canvas edge; it grows closer to its specified growth rate, so becomes larger than the other yellow apples.

The conflict between the yellow apple and its surroundings may be partly resolved by the canvas deforming out of the plane. Figure 30 (plate 6) shows the result of the same growth rules as before, but now the canvas is allowed to buckle and deform in three dimensions. The yellow apples have formed bulges or swellings on the canvas, because by bulging out they can grow larger while not invading their neighboring regions as much. This outward bulging arises automatically from the forces generated by the yellow apples trying to expand faster than their surroundings.

By introducing differential growth into the canvas we have brought about a new form of reinforcement and competition. Reinforcement arises because regions that grow faster than others boost their own size through the law of compound interest. But this process is counteracted by the stresses it generates in neighboring regions, leading to competition for space. The overall result is that the canvas is deformed in various ways.

In our deforming Cézanne, the colors in the finished painting modified its growth. Now imagine the same sort of thing happening while the painting is in progress. A touch of yellow would make one region grow faster; a patch of red applied over the yellow later on might make it slow down. Regions would continually push and pull against each other in various ways as the painting progresses, constantly reinforcing themselves while also competing with each other for space. In addition to creating a pattern of colors, such magic paints would lead to the canvas being deformed in all sorts of ways. This process would in turn modify the context

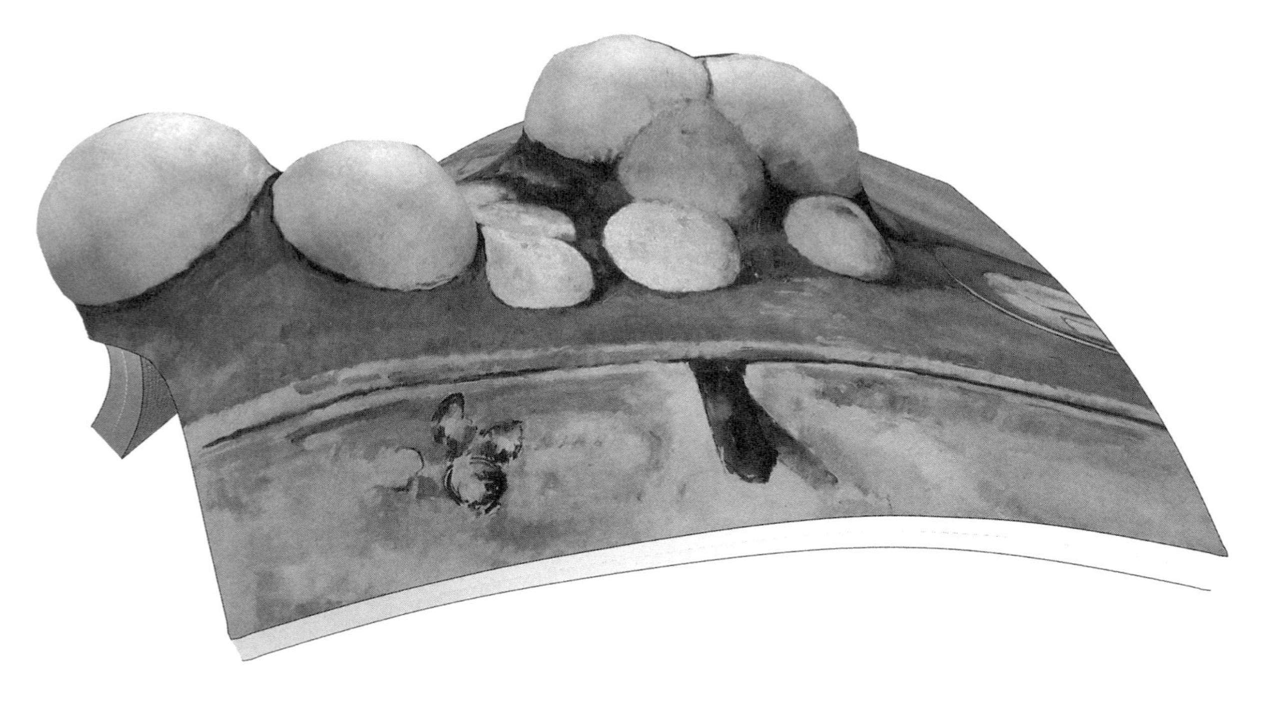

Figure 30. Deformed version of Cézanne's *Apples and Biscuits* when the canvas is allowed to buckle in three dimensions. See plate 6.

of the canvas, its arrangement of colors in space, and then influence the next steps in the process.

Regulatory proteins act like magic paints—they not only generate patterns, but can also influence the way a plant or animal grows. They do this by regulating the genes involved in the organism's physical construction. Recall that the genome contains many thousands of genes. Not all of these genes code for regulatory proteins. Many of them code for proteins that influence the physical properties of cells, including their growth, shape, strength, and adhesion. By affecting the activity of these genes, regulatory proteins can influence the organism's growth and shape. This explains why mutations that disrupt the genes coding for these regulatory proteins can lead to changes in an organism's physical appearance, like a fly with missing segments or an extra pair of wings.

The growth of the snapdragon (*Antirrhinum*) flower provides a good illustration of how deformations can arise through gene action. On the left side of figure 31 is a simplified diagram of this flower at an early stage of development, when the bud is about 1 mm wide. At this stage, the petals form a small cylindrical tube topped by five lobes. Think of this as the shape of a canvas at the start of a painting. This canvas already has various regions depicted on its five segments, reflecting some of the patterns of gene activity that have been established at this stage. For example, regulatory proteins called CYC and DICH are specifically produced in the two darkly colored segments. We can use the regulatory proteins in the flower bud as magic paints to influence the growth of the canvas in a computer. As each region tries to grow in a particular way, the computer simulates the way it pulls and pushes against other regions, causing the canvas to deform. The paints can influence not only growth rates but also the direction of growth. A gradation in color, for example, might allow growth to be oriented preferentially along the direction of the gradient.

The result of these various interactions is illustrated by the snapdragon flower in the middle of the figure 31. This flower is actually much bigger in relation to the starting canvas to its left (about five hundred times the area), but it has been drawn at a reasonable size for ease of comparison. The shape of the final flower arose through genes interacting with each other and influencing the local growth properties of the canvas. Not only does the final shape look like a snapdragon, but its pattern of growth matches how various regions of the flower are known to grow and develop.

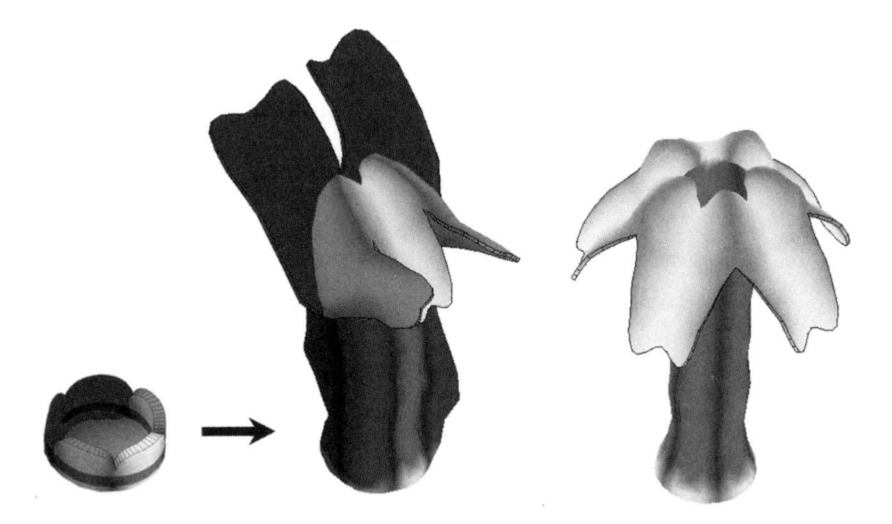

Figure 31. Computer simulated growth of a normal (*left flower*) and mutant (*right flower*) snapdragon.

And if we were to remove some of the regulatory proteins from the developing bud, such as those in the black region, a different flower shape would grow, as shown on the right of figure 31. This resembles a mutant form of the snapdragon which is not able to make the CYC and DICH regulatory proteins.

I have emphasized growth, but the same principles also apply to its opposite, shrinkage. Some of our magic paints can lead to cells dying, the equivalent of shrinking or removing parts of the canvas. Growth and death are highly interconnected and can play off each other in several ways. Consider the branching pattern of a tree. As a young tree grows, new branches are produced; these in turn produce more branches, which themselves produce even more branches. Such a system produces an ever increasing number of branches and twigs through the law of compound interest. This could eventually lead to collisions between branches, or to many useless twigs in the center of the tree that get very little light but nevertheless require resources. One way around this problem is for twigs or branches to die and fall off the tree. If the tree sheds those twigs that are least effective in harvesting light energy through their leaves, then it prunes itself to produce a more effective canopy. Figure 32 illustrates a

Figure 32. Tree generated by a computer program based on reinforcement through branching and competition for light.

tree that was generated by a computer program based on this idea. This program involves reinforcement through the tendency of branches to proliferate. But this growth is counteracted by competition for limited light and space, bringing about death. The overall result is a tree form that matches itself by avoiding internal collisions between branches, while also matching its environment by only keeping those branches that are well lit.

Cell death can have a similar overall effect in animals. The developing nervous system provides a good example. In this case, the issue concerns not absorbing light, but neurons matching the target cells they innervate. During development, our nervous system proliferates to produce more neurons than are needed in the adult. Depending on the particular brain region, 20 to 80 percent of the neural cells that are produced will eventually degenerate. Whether a neuron survives or not can be influenced by whether it receives particular signaling molecules, called neurotrophins. Neurotrophins are survival factors that help keep the neuron alive. Without

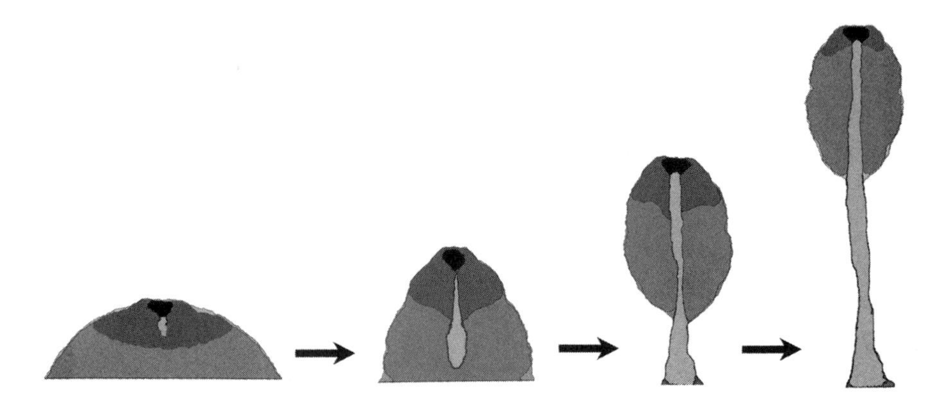

Figure 33. Computer-simulated shape changes in *Dictyostelium* involving cell rearrangements and signaling.

an adequate supply of them, a neuron may die (through a process called apoptosis). Neurotrophins are supplied by neural target cells; if a neuron manages to make effective contact with a target cell, it receives plenty of neurotrophin, allowing it to survive. If it does not make good contact with a target cell, it will likely die. Neurons compete for neurotrophins much as the twigs on a tree compete for light. And the overall result of this competition is that some neurons are pruned back, leaving only those that formed good contacts with their targets.

As well as growth and death, deformations during animal development can arise in still another way—through cell movement and rearrangement. (This process is not as relevant for plants because their cells are held in fixed relative positions by their walls.) The slime mold *Dictyostelium* provides a good illustration of the process. At a certain point in its life cycle this organism comprises a clump of cells that can move and slide over each other (fig. 33, *left*). The clump is subdivided into various regions, indicated by the different shades of gray, with each containing cells that express distinct regulatory proteins. The regulatory proteins influence how cells signal, move, and adhere to each other. As the cells jiggle and slide, with groups associating and reinforcing their movements while also competing for space, the shape of the clump is transformed within a few hours into a stalk with ball at its tip (fig. 33, *right*). Reinforcement and competition are again involved, but now in the form of cell adhesion and movement.

The Three-Dimensional Canvas

A growing and deforming multicolored canvas gives an overall sense of what takes place during development. But a canvas is two dimensional while plants and animals have a complex three-dimensional structure. George Stubbs illustrated the three-dimensional anatomy of horses when drawing their internal structure. He suspended a horse carcass in a farmhouse and peeled off the various layers of the animal, month by month, first drawing the outer muscles (fig. 34) and finally the skeleton. It was painstaking work, conducted in what must have been an unbearable stench. Stubbs published his results in a monumental book, *Anatomy of the Horse*, the best anatomical description of the animal of its time.

Such complex anatomies arise through the various processes we have already covered. Now, however, you have to imagine them taking place on a three-dimensional rather than two-dimensional canvas. To get an idea of what these three-dimensional deformations involve, look at figure 35, which displays some of the key regions in a typical vertebrate embryo. At this stage, the initially spherical egg has already been transformed into a three-dimensional, curved embryo. As we saw for the developing fly, the embryo has become richly depicted with themes and variations indicated by the differently shaded regions. Let's see what happens to some of these regions.

Running along the length of the embryo is a series of repeating units, called somites. The somites are formed in sequence, beginning toward the head end, and involve various genes being switched on repeatedly, much as we saw for the repeating fruit colors in the developing fly. Variations are also superimposed on this repetitive theme, involving additional regulatory proteins that vary from head to tail. Some regions of the somites express regulatory proteins that lead them to proliferate and form the repeating units of the backbone (vertebrae). Other cells expressing a different combination of regulatory proteins proliferate and migrate out to the limbs and switch on genes needed for muscle development. Much of the muscular and skeletal anatomy depicted by Stubbs is the product of somites.

Figure 34. Drawing from *Anatomy of the Horse*, George Stubbs, 1766.

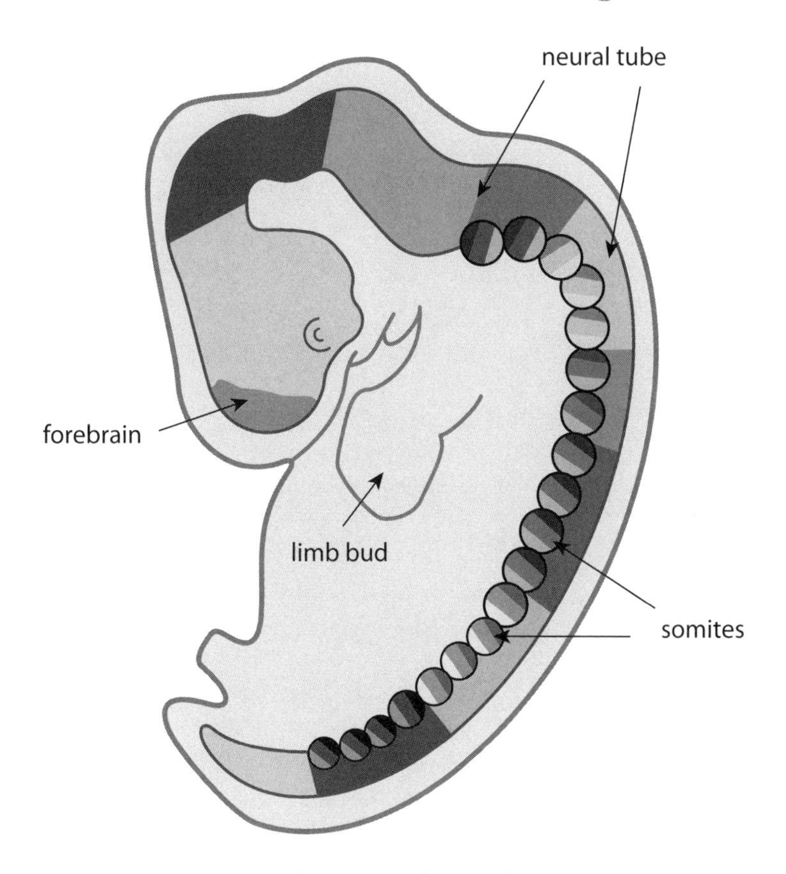

Figure 35. Schematic vertebrate embryo.

Another key structure running along the length of the embryo is a long hollow tube called the neural tube (fig. 35). The wall of the neural tube is made of a single layer of cells. Like the somites, the neural tube is subdivided along the length of the embryo into distinctive regions. Much of the neural tube running along the back proliferates to form various elements of the spinal cord and nervous system. The region of the neural tube near the head forms the brain: it grows and balloons out to form a pair of hemispheres. The wall of these hemispheres is initially one cell thick, but as growth and cell division proceeds, the cells start climbing over each other to form the cerebral cortex, a sheet that is several cell layers thick. The developing cortex is also divided over its surface into regions that express different regulatory proteins. For example, one regulatory

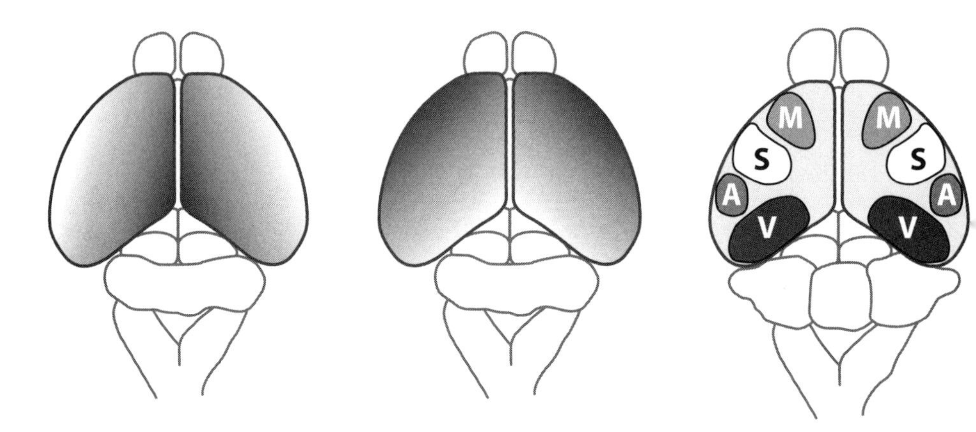

Figure 36. Developing cerebral cortex, showing areas of key regulatory proteins (*left* and *middle*), and the resulting functional regions involved in motion (M), bodily sensation (S), audition (A), and vision (V) (*right*).

protein is present at higher levels toward the central and back end of the cortex (fig. 36, *left*). By contrast, another regulatory protein is expressed more toward the side and front end (fig. 36, *middle*). These regulatory proteins, together with associated signaling molecules, lead to regions with distinct functions forming in the cortex, such as those involved in motion, sensation, audition, and vision (fig. 36, *right*). Mice that lack one or more of these key regulatory proteins develop brains in which the relative size of these regions has changed.

The anatomy that Stubbs depicts arises through recurrent cell patterning, growth, death, and rearrangements of a three-dimensional canvas based on the principles already described.

A Common Recipe

In the previous chapter we saw how a double feedback loop between reinforcement and competition lies at the heart of development. Molecules, such as regulatory proteins, can boost their own levels or actions while also bringing about their own inhibition. This feedback loop is fueled by collisions between large numbers of molecules (population variation), balanced by a degree of molecular stability (persistence). In this chapter we have seen how some additional ingredients also play a key part. One is the interaction and cooperation of molecules or cells in close proximity. This cooperation between multiple components introduces an enormous range of combinatorial possibilities, a vast developmental space of molecular and cellular combinations. Embryos take an odyssey through this immense space by recurrent application of our double feedback loop, constantly modifying their own context and thus setting the scene that moves them onto the next step of the journey. Throughout this process the embryo grows, deforms, and becomes patterned at a range of scales.

This set of ingredients is very similar to what we encountered for evolution in chapters 1 and 2. There we also found a double feedback loop between reinforcement and competition, fueled by population variation and persistence. We also saw that cooperation played a key part, leading to integrated units, such as genes or individuals. In bringing components together cooperation also created a vast space of possibilities: the enormous range of different DNA sequence combinations. Populations were driven through this hyperspace by the ever-thirsty double feedback loop, which recurrently changed the context, and thus spurred the populations on their journey. Like development, evolution is propelled forwards by its past rather than being pulled along by its future.

We can now see how evolution and development involve a similar interplay between our seven principles of population variation, persistence, reinforcement, competition, cooperation, combinatorial richness, and recurrence. Evolution and development are different manifestations of what I have been calling life's creative recipe. For evolution, the recipe involves populations of individuals in an environment, and leads to various life forms arising over many generations. For development, the recipe involves populations of molecules and cells within the same individual, and leads to the emergence of an adult within a single generation. The processes are different in many ways, yet we can recognize a common form behind these different guises.

Of course I have presented evolution and development in a way that emphasizes these commonalities. My reason for taking this approach is that it helps us appreciate the essence of each process. If we consider evolution and development in isolation, we might define a collection of principles in each case, but would probably not arrive at the formulation I describe. Indeed, neither of these processes is normally presented through the seven interacting principles I give. But by stepping back and making comparisons between these processes we gain a view of what is general and what is particular to each case, helping us to understand the fundamentals of how they take place. In later chapters, we will see how this same approach also helps us look at other transformations, such as learning and cultural change.

Notwithstanding their basic similarity in form, we have also seen a striking difference between evolution and development. The path an embryo takes through developmental space forms a repeating loop, with embryos traversing basically the same regions of space every generation. This is in contrast to evolution where the paths taken by species through genetic space are more open-ended and do not have a cyclical character. Where does this difference come from? To answer this question we need to understand how the paths of development arose in the first place. We will turn from looking at relationships of form to those of history.

FIVE

History in the Making

Bashford Dean had two passions in life. One was studying the development and evolution of fish, through which he became professor of vertebrate zoology at Columbia University in 1904, at the age of thirty-seven. The other passion was a fascination with arms and armor. This was inspired in early childhood, when he saw a beautiful European helmet in the house of a family friend. He was so taken with the helmet that he immediately studied it inside and out for a long time while sitting on his friend's porch. His interest in armor increased over the years, and in 1906 he became an honorary curator of arms and armor at the New York Metropolitan Museum of Art. He eventually got permission to retire from his active duties in zoology at Columbia so that he could fully devote himself to making the collection of arms and armor in the Metropolitan Museum one of the finest in the world.

In transferring his interests from fish to armor Bashford Dean carried his biological past with him. This is illustrated by a series of diagrams where he depicted the history of various armaments, such as helmets or shields, much as one might illustrate the evolution of organisms (fig. 37). His diagram of helmets shows a simple ancestral round helmet at the bottom. From this primitive form, various lineages emerge, some leading to the highly elaborate enclosed helmet with a visor, and others leading to various innovations, dead ends, or reversions to simpler forms.

Although such diagrams are helpful in organizing objects and understanding their relationships, they can also be misleading. They give the impression of one object being transformed into another, as if a helmet is

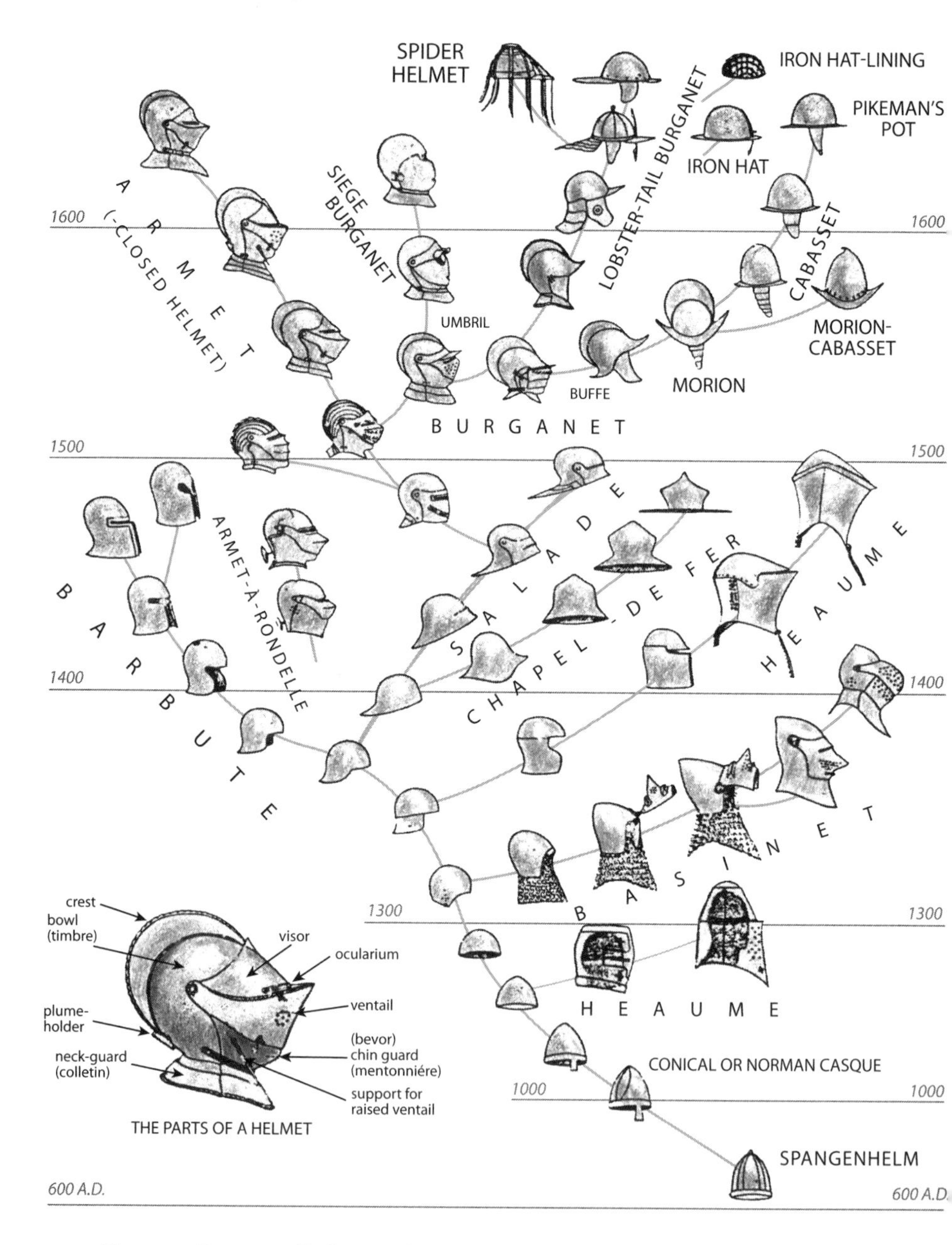

Figure 37. Diagram of helmet evolution, Bashford Dean, 1915.

directly modified to give rise to the next in the series. But this is not the way helmets change over time. The helmet is not transformed; it is the way that people make helmets that changes. The earliest, relatively crude helmets would have been constructed in a few basic steps. Then, as helmet makers grew more experienced, and as the requirements for warfare changed, the helmets became more elaborate and sophisticated. The history of helmets is a history of making.

We can see two interwoven processes in the story of helmets. One is the change in design of helmets over the centuries, the evolution that Bashford Dean depicts. The other is the process of making, which repeats itself many times each generation as the helmet maker churns out numerous copies of the same helmet. These two processes are connected; the actions of the helmet maker depend on the stage in history, while the history of helmets depends on how helmet makers modified their actions over time.

The same considerations apply to the evolution of multicellular organisms. Although we commonly portray evolution as a branching tree along which one type of organism seems to transform directly into another, it is not individual organisms themselves that change, but the way they develop from fertilized eggs. The earliest products of development may have been relatively crude affairs, simple clumps of cells equivalent to the earliest helmets. These simple multicellular structures first arose a billion or so years ago. More elaborate forms of development then evolved, eventually leading to the variety of organisms we see today. For each living multicellular species we see a similar developmental process repeating itself every generation, with multiple organisms of essentially the same kind continually churned out. Yet this repetitive process is not completely fixed but changes over time, through evolution. Development and evolution are closely connected in history. The path of development depends on the evolutionary past, while the path of evolution depends on how development is modified over time.

In previous chapters we saw how evolution and development are based on the same creative recipe. They share a similarity in form. In this chapter I want to look at another type of relationship: the connection of evolution and development in history. As a first step, let's see where the basic ingredients for the recipe of development came from.

Unicellular Beginnings

Development involves the transformation of one cell into a multicellular individual. Yet many of its underlying principles are found in unicellular relatives. Consider the statistical behavior of molecules. The ability of a unicellular organism to perform metabolism or import molecules from its surroundings depends on continual motion that is chaotic at the level of the individual molecules. Regular and predictable metabolism depends on statistical processes, such as diffusion, that involve populations of molecular events. Without the physical process of diffusion there would be no life of any kind. Single-celled organisms are experts at playing the game of diffusion. When two yeast cells of a different mating type need to come together for sexual reproduction, they each produce signaling molecules, called pheromones. These molecules are exported from the cell and enter the environment where they diffuse to stimulate cells of the opposite mating type. Because diffusion involves a fall in concentration with distance, it allows organisms to acquire spatial information—how near or far the pheromone source is. If the concentration of the relevant pheromone is high, the chances are that a cell of the other mating type is nearby. Sexual reproduction is just one example of the many situations in which unicellular organisms use signaling molecules and diffusion to communicate. The principle of population variation applied to development is not something that multicellular organisms invented from scratch; instead, it was exploited a long time ago by unicellular ancestors.

The same is true for the principle of persistence applied to development. Too much diffusion can cause problems for a unicellular creature. If a cell is not surrounded by a membrane that prevents molecules from diffusing into the surroundings, the cell could not persist as an individual. Persistence also operates from one generation to the next. As a single-celled organism divides, its DNA gets copied and the cell contents are shared among the daughter cells. These contents include regulatory proteins, allowing the daughter cells to re-create the pattern of gene activities and proteins of their parent. This is similar to the way patterns of gene activity are transmitted from cell to cell as the fertilized egg of a multicellular organism divides. The same process of cell division that provides hereditary persistence between generations in unicellular organisms helps to maintain continuity and persistence between cells within the same organism during development.

The principles of reinforcement, competition, cooperation, and combinatorial richness as applied to development also have antecedents in the unicellular world. All single-celled organisms regulate their genes in response to changing circumstances. When you drink a pint of milk, a bacterium that lives in your gut, called *Escherichia coli*, modifies the activity of its genes so that it produces more of a protein needed to digest the lactose. A gene coding for this lactose-digesting protein is switched on as large amounts of lactose in the milk modify regulatory proteins in the bacterium. Some of these regulatory proteins activate genes while others inhibit gene activity. Moreover, like the genes of multicellular organisms, those of single-celled creatures have regulatory regions to which many proteins can bind. The extent to which a gene is switched on or off depends on the way these different regulatory proteins interact and cooperate. This makes a large number of different regulatory combinations possible, allowing the cell to respond to the challenges from its environment in many different ways.

Much as many of the tools needed for helmet making were available before the first helmets appeared, many of the elements needed for development were already present in single-celled creatures before multicellular organisms evolved. How did the various ingredients in these unicellular forms come together to provide the recipe for development? Although we were not at hand to witness these events directly, we can present scenarios for what might have happened.

Moving up a Scale

Most of the ocean is in perpetual darkness, but near its surface there is sufficient light to support a thriving community of microscopic plants. These unicellular algae form a vast lawn of marine plankton. There are, however, hazards to this way of life. Plankton is helplessly swept along by currents and turbulent seas. If the microscopic plants are carried down to the lower depths, light levels rapidly fall and they may die in the vast darkness, unless they can swim or float back to the surface. Closer to the shoreline, the shallowness of the seas prevents the fall to gloom. But there is no guarantee that a plant living near the coast might not be swept away into the immense ocean, unless it attaches itself to a rock. Some species of unicellular algae fasten onto the seabed, allowing them to live permanently near the coast. These cells often have two different ends—one is specialized for holding onto the rock and the other is specialized for harvesting energy from sunlight. With this arrangement, they embody an aspect of their surroundings;the interface between opaque, solid rock and transparent, fluid seawater.

With this way of life, individuals that are able to extend upwards and grow away from the rock will gain more light. Their chances for survival may be better than those of their overshadowed neighbors, and they therefore may be favored by natural selection. One means of achieving this growth would be for the cells themselves to become bigger. But there are limits to how large a single-celled organism can grow. As size increases, it becomes more difficult for processes to be coordinated across a large volume of cytoplasm. A partial solution to this difficulty is the inclusion of many nuclei in the cell; this is often the case in the largest single-celled creatures. Still, continual mixing of cytoplasm can make it difficult to establish patterns of gene activity.

Another solution to these problems is for cells to adhere to one another after division, allowing the organism to grow into a multicellular individual. With several cells, each with their own nucleus, the possibility arises of switching different genes on in different cells. Cells near the rock face might switch on genes enabling attachment, while cells at the exposed water end could switch on genes involved in photosynthesis. The evolution of such specialized cells does not require a completely new set of mechanisms. It could arise by the combination of ingredients that were already present in unicellular organisms. Principles that apply to single cells can apply to several cells that stay in proximity. Then the interface between rock and sea can be dealt with in a new way, through differences between cells rather than within them. The alga has captured or embodied a spatial feature of its environment, the distinction between rock and the sea around it, through its organization of cell types. It has carved up the world by carving up itself.

Our primitive alga is able to achieve this organization by following a relatively simple, circuitous path in developmental space. The initial cell of the alga starts with a particular combination of regulatory proteins, corresponding to one location in developmental space. This context provides the molecular conditions that drive the cell to divide and produce a few more cell types in the early embryo, some better suited to gripping the rock and others to harvesting light. The embryo has been propelled to a new location in developmental space. This in turn provides the context that drives the next step. Eventually, cells are also set aside for the formation of the reproductive cells that give rise to the next generation. In this way, the alga forges a looped path through developmental space, governed by a set of molecular and cellular principles that were already operating in its unicellular ancestors.

This scenario demonstrates how ingredients already present in the unicellular world may have come together during evolution to provide a basic recipe for development. Once this recipe was in place, more elaborate forms of development could evolve. Our simple multicellular algae stuck to the rock compete for light, so forms that can proliferate and grow taller during their life cycle may be favored by natural selection. In growing taller, additional issues crop up, like the stresses brought about by the plant being wrenched by currents, or the problem of cells at the rock face staying alive as they lie further from the photosynthetic cells. Additional cell specialization may be helpful, such as a collar of stronger cells that prevent the plant from being torn off by currents, or a transportation system for moving sugars from the tip to the base of the plant. No fundamentally new mechanism is needed for such specializations to evolve. They can arise from the same process of patterning being repeated as the organism develops. By recurrently building pattern upon pattern as the organism grows, a range of specializations can arise in an organized fashion. The overall result is that our looped path through developmental space has become extended and modified.

I have offered a simplified scenario for how our seven principles applied to development may have been put together during the evolution of multicellular plants. A similar story can be told for animals. Many single-celled animals live successfully by consuming other organisms. But there are some advantages for animals being larger, including the ability to swallow other creatures, and avoid being swallowed in turn. Increasing size by becoming multicellular has the added benefit of allowing different cells to acquire specialized jobs; some cells might be dedicated to eating and others to digesting food. And as the animals living at a greater scale encounter further challenges, such as moving around effectively or coordinating different body parts, further cell types and arrangements could arise through recurrent patterning that deal with these challenges.

Greater size and complexity does not come without some costs. For example, it delays reproduction because the organism needs more time to grow to its mature form. The benefits of size therefore need to be set against the costs of an increased generation time. Trade-offs like this abound in the living world because improving in one way can often be at the expense of growing worse in another. For this reason multicellular organisms have not replaced unicellular ones: single-celled creatures like bacteria continue to vastly outnumber their many-celled relatives. As Stephen Jay Gould has pointed out, we still live in the "Age of Bacteria," a period which has lasted about 3.5 billion years. Rather than an overall progression of life to ever greater size and complexity, ecosystems contain many different forms that coexist, each capturing relationships at various scales, from the microscopic to the macroscopic.

Zooming and Growing

When we look at a seaweed of today, such as bladder wrack (*Fucus vesiculosus*, fig. 38), we see that it is adapted to its environment at many scales. Viewed as a whole organism, it forks and branches over distances of tens of centimetres, an adaptation that enables the plant to harvest large amounts of energy from sunlight. It produces a rootlike structure at the base, which anchors the plant to rock. This holdfast is connected to the rest of the plant by a strong stem that resists the continual buffeting of the sea. Looking a little closer, we see that the main body of the alga is flattened like a sheet, providing a high surface area to harvest light energy. The sheet's midrib gives it extra strength to resist the motions of the sea. Some regions are also swollen to form air-bladders that provide buoyancy. If we zoom further into the plant we see individual cells, each about one tenth of a millimetre across. There are various cell types that are specialized in different ways, such as those in the main sheet that carry out photosynthesis, or those in the holdfast that attach the plant to rock.

Zooming in further, we see that each cell contains a series of exquisitely organized, small compartments, each about one thousandth of a millimeter long. These include chloroplasts specialized for carrying out photosynthesis, and mitochondria, the cell's powerhouses. Zooming even further into the chloroplast compartment, beyond the limits of the light

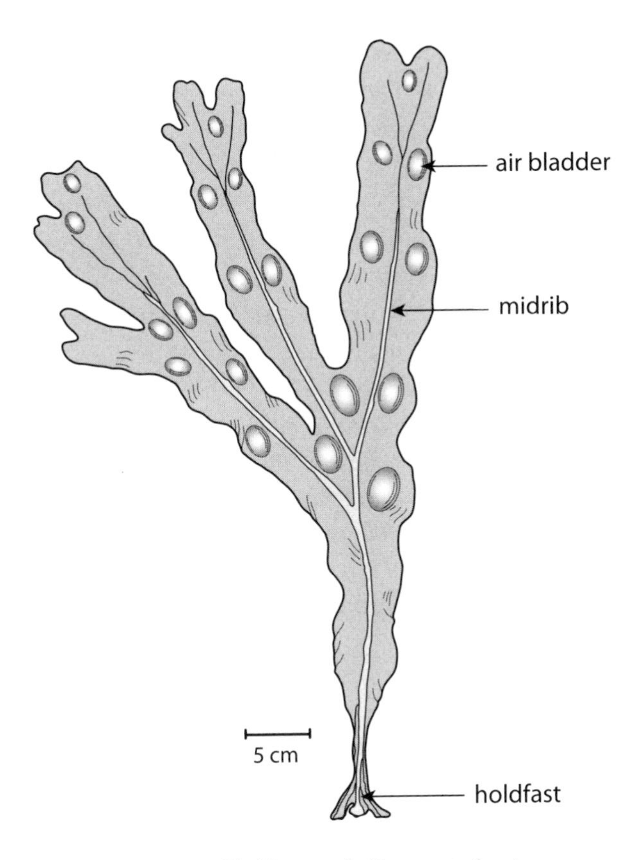

Figure 38. Bladder wrack (*Fucus vesiculosus*).

microscope, we see swarms of tiny bodies in a variety of shapes, each about one hundredth of a thousandth of a millimeter across. These are the proteins that promote chemical reactions in the cell, such as the Rubisco protein mentioned earlier (chapter 2 on page 45), which catalyzes the fixation of carbon dioxide.

All of these levels of organization represent the many different scales at which the sea weed captures relationships in its environment. The branching architecture of the plant reflects relationships such as the way sunlight falls on the plant, and the effects of sea currents and gravity. The shape of the Rubisco protein matches the spatial relationships in a molecule of carbon dioxide—the arrangement of atoms and electrical charges over the surface of the molecule. At every scale, the adaptations of a plant involve capturing relationships in space and time in the environment it interacts with.

The same applies to unicellular creatures. The Rubisco protein from bladder wrack has a similar shape to Rubisco from single-celled algae. In both cases it operates at the molecular scale to fix carbon dioxide. But some aspects of the environment are closed to unicellular organisms. The wrenching action of ocean currents has little direct relevance for a single cell. If you were a fraction of a millimeter long you could easily hold onto a rock no matter how powerfully the waves crashed around you. Water at that scale would feel viscous, like molasses feels at our scale. So, even though unicellular life is rich and varied, it is oblivious to some of the relationships in the larger scale world. Larger multicellular organisms can operate in this greater world. And they are able to capture many of its relationships because of their cell types and organization at multiple scales.

They achieve this organization by traversing scales in the opposite direction to zooming in: through growth and proliferation. The bladder wrack starts as a fertilized egg that drifts about until it anchors to the sea bed. It then divides into two cells with distinctive properties. One of the cells goes through further rounds of division and patterning to form the fronds of the plant. The other undergoes growth, division, and differentiation to form the holdfast. The plant climbs up its various scales of organization through recurrent growth and patterning.

A Recipe within a Recipe

During one of his trips to Europe, the armor enthusiast Bashford Dean came across an old box in the corner of an ancient attic. The box belonged to an armorer of about 1600 and contained parts of unfinished gauntlets. Bashford Dean remembered the incident:

> It gave me a curious feeling to take in my hands these ancient objects which seemed only yesterday to have been put in the box by their maker. I had the strong impression that if I should go through the old door near by, I would by some "Alice in Wonderland" wizardry, pass into the sixteenth century and find in the next room a veritable armorer at his table by the low window.

In this chapter we have also been transported into the past. We have seen how evolution and development are connected in history. Many elements of development were already lurking in our unicellular ancestors that existed billions of years ago. These ingredients eventually came together in a particular way during the evolution of many-celled creatures. The earliest multicellular forms exhibited a limited degree of recurrence during development, displaying a simple organization of two or three cell types. These changes represented relatively simple, circuitous paths through developmental space. But as further levels of recurrence evolved and the paths became more complex and extended, creatures could explore and capture relationships in their surroundings at many scales, resulting in new ways of surviving and reproducing effectively.

Instead of being fixed, the circuitous paths of development are continually shifting and deforming on an evolutionary time scale. This is because journeys through genetic and developmental space are connected. As populations shift through genetic space during evolution, genes that influence developmental paths may change, allowing organisms to capture new relationships in their environment. From this grander perspective, development does not always follow precisely the same path and return to exactly the same place each generation. The paths gradually shift, so that over evolutionary time they form a series of spirals rather than strict loops. If we trace the chicken and egg far enough back in time, we do not find chickens and their eggs but instead we find their early ancestors, organisms comprising just a few types of cell. Going even further back in time we would not come across developmental paths at all, but simply the reproductive cycle of unicellular creatures.

Like nested dolls, evolution and development exhibit a double relationship. On the one hand, development is historically embedded in evolution; it arose through, and is contained within, the evolutionary process. On the other hand, development has a similar form to that of its evolutionary parent; they are based on the same creative recipe. Even though one operates over many individuals and generations, and the other within a single individual and generation, we find common, fundamental principles at play in both cases.

This is not the end of the story, however. For just as evolution spawned development, we will see that development led to yet another process following the same recipe: the process of learning. To pursue this next transition we must first get a better idea of how learning works. Rather than immediately trying to understand how complex creatures like dogs or humans learn, a good stepping stone is to first look at the behavior of more humble organisms.

SIX

Humble Responses

ONE OF THE MOST CURIOUS PAINTINGS of the Renaissance is a careful depiction of a weedy patch of ground by Albrecht Dürer (fig. 39). Dürer extracts design and harmony from an apparently random collection of weeds and grasses that we would normally not think twice to look at. By choosing such a mundane subject he is able to convey his artistry in a pure form, uncontaminated by conventional distractions. In a similar way, scientists often choose to study humble subjects when trying to get at the essence of a problem. Studying relatively simple systems avoids unnecessary complications and distractions, and can allow deeper insights to be obtained. This is particularly true when we are trying to understand something as problematic as our ability to learn. Human reactions are so complex and intertwined with our emotions that they can be difficult to interpret objectively. It sometimes helps to step back and consider how more modest creatures, like bacteria, weeds, or slugs, deal with the challenges they face.

In the previous chapter we looked at how organisms deal with spatial challenges, such as the distinction between rock and sea. Through development they acquire distinctions in their anatomy that allow them to capture some of the heterogeneity in their surroundings. In this chapter I want to start looking at the time dimension, at how organisms deal with change. If the environment was always constant, an organism would never have to modify its behavior according to circumstance. In this situation, processes like learning would have much less significance. To arrive at the roots of learning we first have to understand the basic ways in

Figure 39. *Large Piece of Turf*, Albrecht Dürer, 1503.

which organisms respond to changes in their surroundings. We will consider a range of responses in microbes, plants, and animals so that we can begin to discern some common as well as distinctive elements. Plants, for example, do not have neurons, but they can nevertheless capture quite complex patterns of change in their surroundings. Once we have established the basic biological mechanisms for dealing with change, we can then look at how these various mechanisms are brought together to give rise to learning, our third instance of life's creative recipe.

Making Adjustments

We live in a mutable, ever changing world. Rather than being inflexible, it pays for organisms to be able to adjust to changes in their circumstances. We have already encountered one way of responding to change: as mentioned in the previous chapter, the bacterium *Escherichia coli* can adjust its levels of gene activity according to the amount of lactose in its surroundings. It has regulatory proteins that can switch genes on or off in response to lactose levels. This response is an adaptation; it allows the bacterium to produce proteins, such as a lactose-digesting enzyme, that help it survive and reproduce. All unicellular organisms have regulatory mechanisms of this kind, allowing them to respond to many factors that may change with time, from the prevalence of particular molecules to temperature or light.

In some cases these responses can be very swift. The single-celled green alga, *Chlamydomonas*, commonly found in ponds, can sense light and move toward it using two tiny filaments, called flagella (fig. 40). Light-sensitive receptor proteins in the *Chlamydomonas* cell are modified when light hits them and this leads to the generation of an electrical signal, which sweeps across the cell. The signal changes the pattern of flagella movements so the cell turns toward the light, just as a swimmer may adjust their stroke to make a turn. By swimming toward light, *Chlamydomonas* ensures that it can stay in bright surroundings, and continue to survive through photosynthesis.

These responses—reacting to lactose or swimming toward light—are adaptations that provide effective methods of dealing with a variable environment. They are called homeostatic mechanisms. With these mecha-

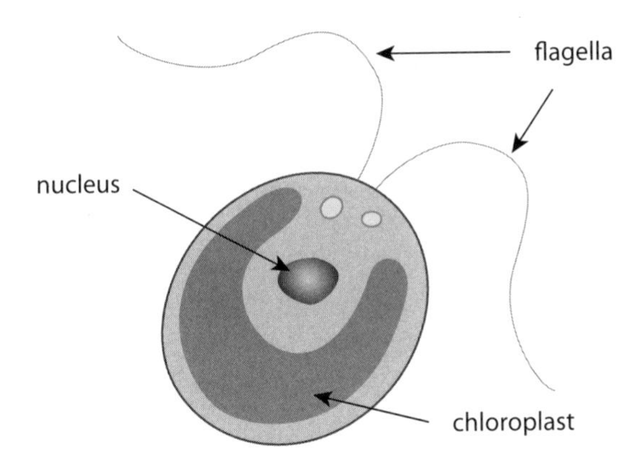

Figure 40. *Chlamydomonas* cell.

nisms, an organism can maintain a well nourished and balanced state in spite of changing conditions. As with other adaptations, they have been selected for through Darwinian evolution. Two different time scales are involved here. First, environmental variations within the lifetime of an individual influence the way it reacts. Second, on an evolutionary time scale, certain reactions have been selected for. Over countless generations, natural selection has favored some reactions over others, so only those that promote survival and reproduction have become established. Consequently *Escherichia coli* regulates its genes when presented with lactose, and *Chlamydomonas* swims toward light—these are responses that have helped individuals contend with the variability in their surroundings in previous generations. They are instinctive responses.

Instinctive responses need not be fixed, however; they may adjust over time. A good example is the swimming behavior of the bacterium *Escherichia coli*. Like *Chlamydomonas*, *Escherichia coli* can swim using flagella, although in this case the flagella act like rotating propellers rather than flexing arms. The bacterium instinctively swims toward areas that are rich in food substances, such as sugars, and away from noxious compounds like acids. It does this through receptor molecules in its membrane that continually sense levels of chemicals in the surroundings. These receptors influence bacterial movement through a strategy similar to that of a child playing the game of "hot and cold"; movements are adjusted according to whether the bacterium is getting closer (hotter) or further away (colder) from the source of interest. The bacterium first heads off in a random direction. If it senses that conditions are improving, with the concentration of sugar increasing, then it continues in the same direction. But if it senses that conditions are getting worse, the bacterium does a tumble, and swims off in another direction selected pretty much at random. Through a series of swims and turns it eventually reaches the source of food.

Escherichia coli can perform in this way because of a remarkable property of its receptors: their sensitivity is gradually modified according to how much they are stimulated. If there is a lot of sugar around the bacterium, its receptors slowly become less sensitive to sugar, so it takes more to trigger them. This means that if sugar is abundant for a long time, the bacterium becomes habituated and no longer responds to the high sugar levels. The bacterium only responds strongly when food levels are changing. If it is swimming toward a source of sugar, it senses the increasing concentrations and keeps going; if it is swimming away from a food source, it senses decreasing concentrations and switches direction. The bacterium does not just respond to the prevailing situation, but compares its sensory input to its recent history of experiences in order to determine whether food availability is improving or not. It captures relationships in time and uses these to determine its path in space.

Natural selection has led to the evolution of many instinctive responses as adaptations to variable environments. These responses capture relationships in the surroundings both in space and time. When someone drinks a glass of milk, the milk passes through space, down into the gut where it floods the surroundings of *Escherichia coli* bacteria with a supply of lactose. As the Earth rotates in space, the way light falls on a pond will change, and so will the direction in which *Chlamydomonas* swims. As *Escherichia coli* propels itself along, it may encounter a rich source of food and swim toward it. Time and space are linked, and many elements of this relationship are captured through instinctive responses.

Given their prevalence in unicellular organisms alive today, instinctive responses must have arisen very early in biological evolution, when the living world comprised only single-celled creatures. But with the advent of complex multicellular organisms about half a billion years ago, these responses were taken to a different scale. Instead of cells reacting in isolation, responses became coordinated among multiple cells in an individual. The plant world provides some striking examples.

Flora's Story

If Dürer's painting of a piece of turf is one of the curiosities of Renaissance art, then Giuseppe Arcimboldo's paintings get the prize for eccentricity. In *The Librarian* (fig. 41, plate 7) he uses books and paper to construct a bearded man complete with an arm and hand. The unusual juxtaposition of objects and meanings gives this painting more affinity with the Surrealist paintings of the twentieth century than with other works of the Renaissance. But perhaps the most famous example of Arcimboldo's work is his depiction of the four seasons (fig. 42, plate 8), where he composes human heads from plant parts for various times of the year. For *Spring* he uses flowers to compose the head; for *Summer* he uses fruits; for *Autumn* he makes up the head with late harvest products like potatoes and grapes; while for *Winter,* he uses a gnarled tree stump. Plants provide ideal material for illustrating the seasons because they undergo such dramatic changes throughout the year.

In parallel with the plant transformations, the heads in Arcimboldo's paintings age from a youth in *Spring* to a wizened old man in *Winter*. But while there are parallels between human aging and seasonal changes in plants, there is also an important difference. We age without reference to our environment. Of course there are factors in our surroundings, such as the food we eat or exposure to the elements, which may promote or delay the onset of wrinkles and sags. But overall, aging occurs relentlessly without respect for our circumstances. By contrast, many of the key transitions in plant life, such as flower production, take place as specific responses to the environment. Plants regularly adjust their form according to the time of year. How do they do this?

One of the key environmental factors that plants can monitor is day length. Some plants respond to the lengthening days following winter, and only initiate flowering when the days get sufficiently long. Sensing day length involves being able to time the interval between two events: sunrise and sunset. To achieve this, plants must be able to monitor the amount of light, and also be able to mark time.

Figure 41. *The Librarian*, Giuseppe Arcimboldo, 1565. See plate 7.

The ability to sense light is prevalent in plants. Light is an important part of their existence, providing the vital energy input for photosynthesis, so plants have evolved numerous ways of sensing the level and quality of light. They achieve this through a variety of receptor proteins that set off trains of reactions when triggered by light; these reactions include the switching on or off of genes. So, the first requirement for measuring day length, the ability to detect and respond to light, is not only to be found in animals but is also a basic feature of plant systems.

Perhaps more surprisingly, plants also have an internal clock. This was first discovered in 1729 by the French astronomer Jean-Jacques d'Ortous de Mairan. He was interested in why mimosa plants fold their leaves at night and open them during the day. To see if this was related to the daily variation in light levels, he placed some plants in a dark cupboard. By peeking in every so often he noticed that the plants continued to open

Figure 42. "The Four Seasons": *Spring* (*top left*), *Summer* (*top right*), *Autumn* (*bottom left*), and *Winter* (*bottom right*), Giuseppe Arcimboldo, 1573. See plate 8.

and close their leaves with a daily rhythm even though they were continuously kept in the dark. Variation in external light levels was not essential for maintaining the rhythm. Instead, some sort of internal process was involved, as if the plant had a built in clock.

It is only within the last few decades that we have come to understand how biological clocks function in plants and animals. For plants, this came about by studying *Arabidopsis thaliana*, a common weed that would not have looked out of place in Dürer's turf. This is the plant equivalent of the fruit fly, with a similar ability to reproduce quickly in the laboratory. Studies on *Arabidopsis* revealed that its internal clock relies on the daily rise and fall of particular regulatory proteins. Recall that regulatory proteins can bind to genes and activate or inhibit them. The regulatory proteins involved in the clock go up and down in concentration every twenty-four hours as the result of a feedback loop. As the concentration of one of the activating regulatory proteins rises in a cell, it starts to switch on the production of an inhibitor protein. The inhibitor then starts to increase and prevents more of the activator from being made. Without much activator present, the inhibitor is no longer produced and this allows the activator protein concentration to start increasing again, completing the loop. It takes about twenty-four hours for the system to run through a full cycle like this. The feedback system is similar to what takes place in Turing's model described in chapter 3, except here we are creating patterns in time rather than space.

There is another key feature of the plant's clock. It is normally set by external events like sunrise, as if a clock is adjusted to a fixed hour every day at dawn. This does not affect the rate at which the clock runs—it still runs through one cycle about every twenty-four hours. But the cycle is normally shifted to coincide with daily environmental fluctuations. (This does not happen when plants are artificially held in continuous dark; the clock then continues without regard to its surroundings.) The shifting takes place because the cycling of regulatory proteins is influenced by transitions between dark and light. The appearance of light in the morning leads to a molecular change in the setting of the clock that persists, at least until the next sunrise when the clock may be reset again. Consequently, the plant clock doesn't time events in a fixed manner, but instead the timing is relative to external events like sunrise.

Let's now return to the problem of sensing day length. Plants accomplish this by integrating their ability to sense light levels with information from their internal clock. As days lengthen in spring, leaves detect light at later and later times relative to sunrise. These later times correspond to later phases of the internal cycle of regulatory proteins. Eventually, light detection at a particularly late phase of the protein cycle triggers the leaves to produce a signaling molecule that travels to the plant's growing tips. The molecule then activates genes that promote flower development, causing floral buds to start forming. The spring flowers we witness every year depend on this integration of sensing light, keeping time, and signaling. Through this process, many plants are able to detect the relative timing of two events, sunrise and sunset, and thus determine the time of year.

In adjusting to the seasons, plants are able to capture relationships in time. But these relationships are also embedded in space. The setting of the plant clock is related to the daily spinning of Earth on its axis. And the seasonal changes in day length reflect Earth's annual rotation around the Sun. Plants essentially coordinate their development with planetary movements. They do this not because of any interest in astronomy, but because of the benefit of flowering at particular times of the year—flowering at an appropriate time is an instinctive response that allows plants to reproduce more effectively.

The story shows how flowering plants are able to respond to temporal and spatial features of their environment, even though they may be made up of many millions of cells. The responses of individual cells are coordinated by signals traveling between them, so the plant is able to respond as a whole. I have described plant responses to day length but this is only one of many responses to their surroundings. Plants also integrate their reactions to factors like temperature and nutrients. When a plant switches to flowering, it is essentially making a decision based on a whole constellation of different environmental factors. The same applies to other responses, such as the shedding of leaves or the germination of seed.

These responses involve changes in a plant that may persist for hours or months. The slightly earlier appearance of the sun in the morning can shift a plant's internal clock of regulatory proteins, a change that persists at least until the next sunrise. It may take exposure to only one or two longer days for a plant to initiate flower formation, but once the flowering system has been triggered it may continue for many weeks, even if the plants are transferred back into short day conditions. The long days led to a persistent change in gene activity. This in turn leads to a lasting modification of the spatial organization of a plant through the production of flowers.

It is because changes can persist that plants are able to respond to a sequence of events in time rather than just single events in isolation. The rising of the sun shifts a plant's internal clock and thus modifies the plant's response to light later in the day. An event in the morning lasts through the evening, allowing one response to build on another, and enabling a plant to detect whether it is a spring or winter day. This can happen because plants are not only responsive to their surroundings, but some responses can change the responsiveness of the plant itself. This is similar to the responsiveness of the swimming bacterium *Escherichia coli*, which adjusts it sensitivity according to how much sugar it recently sensed around it.

Although instinctive plant responses, like the initiation of flowering, might impress us with their elaborate mechanisms, they nevertheless appear slow to us. It may take weeks or even months for their effects to become evident, so we tend to take them for granted as part of the yearly cycle rather than viewing them as sophisticated reactions. We are more struck by responses that take place on a shorter time scale. Quick, visible reactions are rarer in the plants than in animals, but there are nevertheless some informative examples.

The Bite of Venus

If an insect happens to crawl over a leaf of the Venus fly trap (*Dionaea muscipula*), the leaf rapidly snaps shut (fig. 43). The unfortunate insect is then trapped by a cage of interlocking "teeth" at the leaf edges and is slowly digested by secretions from the leaf's inner surface, allowing the plant to harvest valuable nutrients. Charles Darwin was fascinated by the Venus fly trap, calling it "the most wonderful plant in the world." Darwin observed that the closing response of the leaf is activated by touch-sensitive hairs on its inner surface that are triggered when bent by the movement of the insect. Once triggered, the hair cells generate an electrical pulse, first detected in 1873 by Darwin's collaborator, John Burdon-Sanderson. This pulse, known as an action potential, spreads throughout the leaf and is sensed by its cells, causing it to snap shut by a hydraulic mechanism. The touch response of the Venus fly trap is very rapid, occurring in 0.3 seconds—it has to be this fast to deal with the rapid movements of its prey.

Not only does the Venus fly trap demonstrate a rapid response, but these responses can build on each other in several ways. If a trigger hair is stimulated weakly, by only a slight movement, then it generates a small response, which is not enough to fire an electrical pulse through the leaf. The stimulus is said to be below the threshold needed to generate an action potential. However, if several weak stimuli are given in quick succession, they may add up until they reach a sufficient level to set off an electric pulse. Each weak response can persist for enough time for it to be added to the next, as if the hair retains a memory of each event. Also, setting off

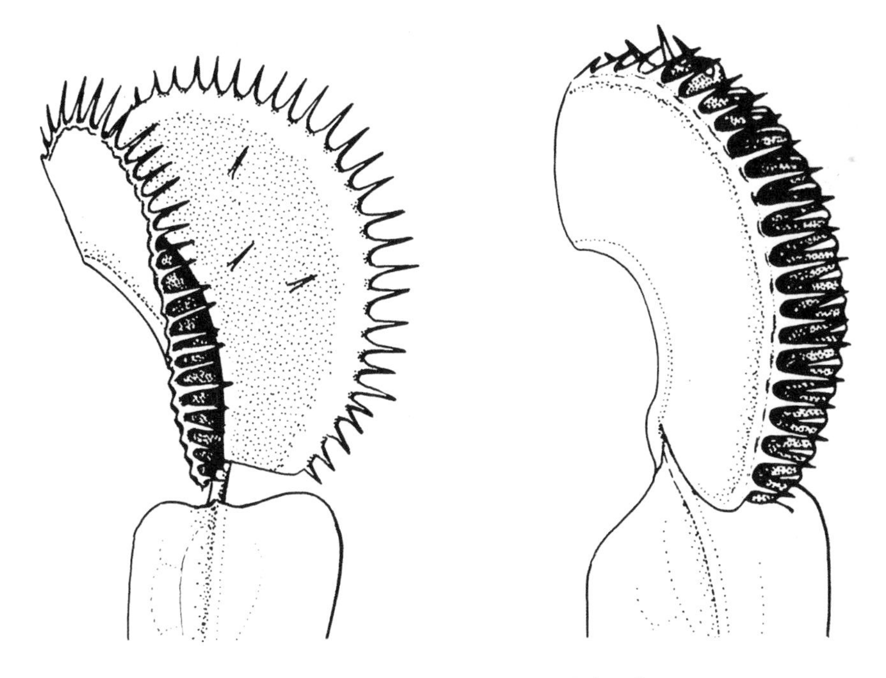

Figure 43. Leaf of the Venus fly trap shown open (*left*) and closed (*right*).

just one electric pulse in the leaf is not enough to trigger shutting. A second pulse has to be fired within 35 seconds of the first one, either by a touch to the same hair or a different one. Again, persistence is involved that allows information from the first pulse to stay around until the second one arrives, enhancing the first one's effect. The plant can then distinguish an insect crawling over a leaf, which is likely to trigger multiple events in close succession, compared with a single event of something brushing against the leaf. As with day length, the plant is basing its responses on a sequence, or a history, of events rather than just an isolated occurrence.

The Venus fly trap has developed structures that allow it to capture relationships in space and time. It can tell from the sequence of stimuli when an insect is likely to be crawling over a leaf. The response is also spatially specific—only the leaf that is being triggered will close. The spatial organization of the plant, the multiple leaves it has generated during development, allows it to respond according to location. Essentially, the plant is able to tell both where and when an insect arrives.

In the case of the Venus fly trap, stimuli can reinforce each other to increase a response, but there are also cases in which repeated stimulation

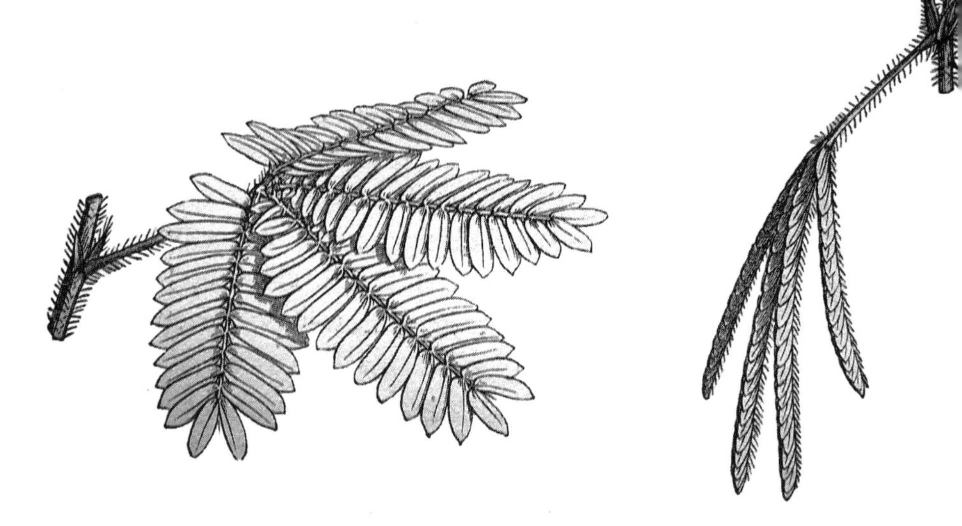

Figure 44. Leaves of *Mimosa pudica* before (*left*) and after (*right*) stimulation.

has the opposite effect, and reduces a response. If you touch the leaflets of *Mimosa pudica* they quickly fold together, closing up the leaf (fig. 44). This is because an electrical signal is triggered by the movement of the leaflets caused by touch. In 1960, when a typhoon hit Tokyo, Hideo Toriyama at the Women's University noticed that after a period of repeated buffeting the local *Mimosa pudica* plants stopped closing up their leaves and kept them open. It was as if the leaves had become habituated to the strong wind and then ignored the continual stimulation. Toriyama later replicated the effect in controlled conditions by placing plants in front of an electric fan. She showed that it took about ten hours for the habituation to build up. Here responses persisted and added up to reduce overall sensitivity, just as we saw for *Escherichia coli* becoming habituated to high sugar levels.

From our anthropocentric view of the world, we tend to think of plants as rather passive organisms. But when we look more closely, we see that they actively respond in all sorts of ways to their environment. Their responses may require fractions of a second or up to several months, but the principles are always the same; in all cases, plants react in a way that is likely to favor survival and reproduction. Moreover, responses are coordinated among many cells through molecular or electrical signals, allowing a harmonized reaction. And in many cases, the responses lead to changes in the plant that may persist, and change the way it later responds. The rising of the sun sets internal molecular clocks, modifying how plants respond later in the day. Triggering a Venus fly trap hair or continuously buffeting a *Mimosa* leaf can change the way these systems react to further stimuli. In these kinds of ways, plants capture sequences of events in time, such as the changing seasons, an insect moving about on a leaf, or the persistence of a storm.

All of these responses depend on plants being spatially organized at a range of scales. The control of flowering time involves plant parts like leaves, stems, and growing tips. The reaction of the Venus fly trap depends on the touch sensitive hair cells embedded in each leaf. As we have seen in previous chapters, these anatomical arrangements arise through the process of development. The fertilized plant egg grows and divides to form roots, stems, and leaves, with their numerous cell types. These developmental patterns are adaptations that allow the plant to survive better in an environment that also exhibits patterns in space and time.

We can also see that responses to the environment can feed back to affect the process of development. The development of flowers may depend on a response to an environmental factor, such as the length of a day. We tend to regard development as a fixed process that happens with little reference to its surroundings. Perhaps we take this view because our own early development, in our mother's womb, occurs in protected circumstances. But the separation between development and responses to the environment is not so sharp. Many instinctive responses can lead to persistent changes in bodily organization, in the pattern of cells and their connections in the organism. These changes may occur while the organism continues to grow and develop. There is no clear line between development and responses to the environment; they are highly interwoven. We have looked at this interaction in plants, but similar principles apply to animals, as we will see.

The Sensible Sea Slug

During his voyage on the *Beagle*, Charles Darwin was captivated by the reflex reactions of a large sea slug that he observed while staying on the Cape Verde Islands. He noted in his journal: "A large *Aplysia* is very com-

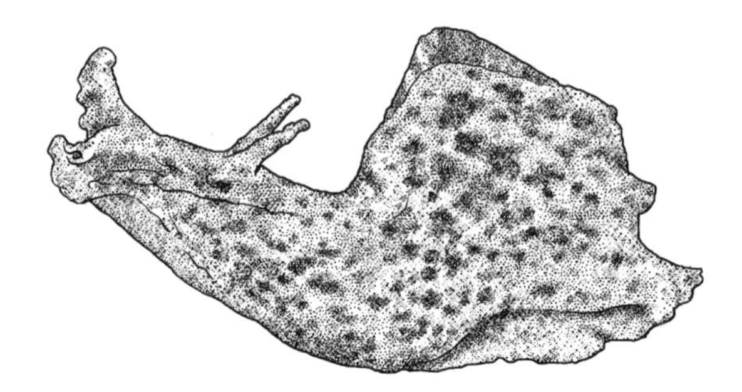

Figure 45. The sea slug *Aplysia californica*.

mon. This sea slug is about five inches long; and is of a dirty yellowish color, veined with purple.... This slug when disturbed, emits a very fine purplish-red fluid, which stains the water for the space of a foot around." For a long time the reactions of this animal remained no more than a zoological curiosity. But things changed in the 1960s, when Eric Kandel at New York University decided to use a Californian variety of this slug, *Aplysia californica* (fig. 45), as a subject for studying the detailed mechanisms of animal responses. Like other animals, the reflex responses of *Aplysia* depend on nerve cells, also known as neurons. The advantage of the sea slug over many other beasts for research is that the neurons are relatively large, making them easy to see and manipulate. There are also not too many of them—the brain of *Aplysia* comprises about twenty thousand neurons, as compared to the one hundred billion or so in our brain. But before going into more details about this system, let's first look at some general features of neurons.

Figure 46 is a simplified diagram of two neurons connected together. You can see that each neuron consists of several regions. There is a central body, containing the nucleus, from which many short branches emerge, called dendrites. Each neuron also includes a long projection, called an axon. The axon ends with some terminal branches which have small knobs or swellings at their end that lie very close to the membrane of another neuron. These junctions are called synapses.

Each neuron can be either resting or firing. For much of the time a neuron may be in the resting state, during which nothing much happens.

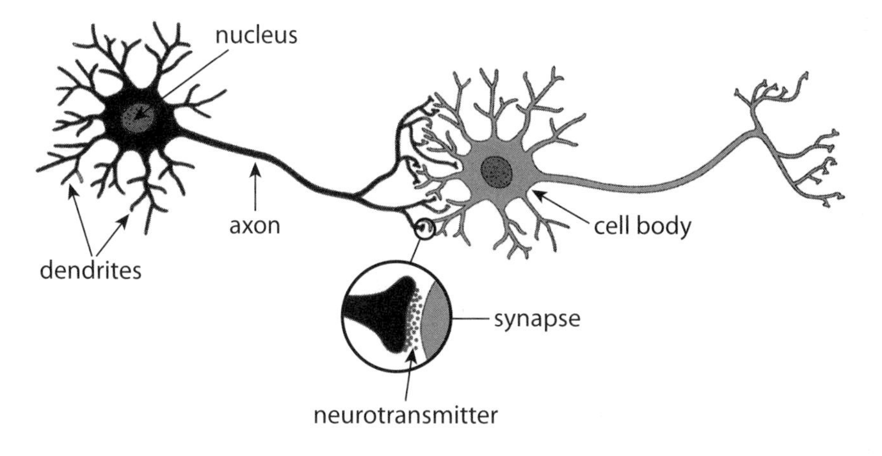

Figure 46. Two neurons.

However, a neuron may switch to the firing state, during which it sends an electrical pulse or action potential from the cell body down the axon. The effect of this on the next neuron depends on what takes place when the pulse arrives at the synapses at the end of the axon. Each synapse acts as a relay station that converts the incoming electrical signal into a chemical signal that travels across the gap between one cell and the next. When an electrical pulse arrives at a synapse, signaling molecules, called neurotransmitters, are released and rapidly diffuse across the gap to trigger receptors in the cell membrane on the other side. Neurotransmitters can act in either an excitatory or inhibitory fashion. In the excitatory case, the neurotransmitter increases the chance that the cell on the other side of the gap will fire; while in the inhibitory case, the neurotransmitter decreases the chance that the next cell will fire. Depending on the types of neurotransmitter released and the receptors on the other side of the gap, a synapse may be stimulatory or inhibitory.

Let's now return to our sea slug *Aplysia*. If you gently touch the slug's siphon—a fleshy spout that expels water—the animal quickly withdraws its gills. This withdrawal reflex is to protect the slug's breathing apparatus from possible damage. Most of the slug's nerve cells are clustered in groups called ganglia. The gill-withdrawal reflex is controlled by one of these ganglia—the abdominal ganglion. This ganglion contains about two

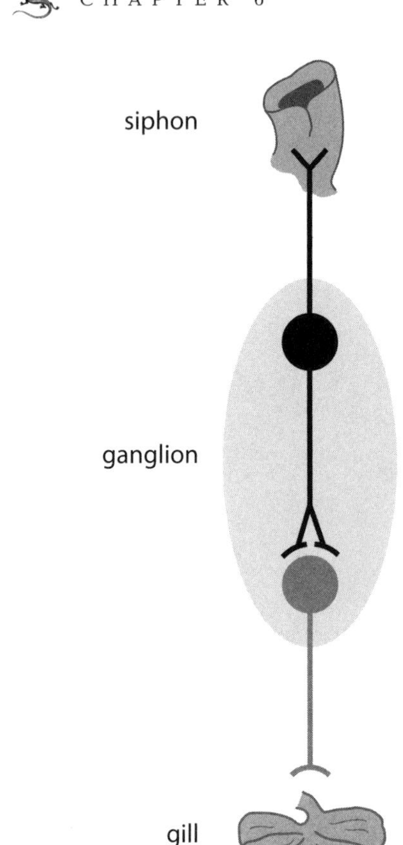

Figure 47. Sensory (black) and motor (gray) neurons connected by synapses in the abdominal ganglion of the sea slug *Aplysia californica*.

thousand nerve cells, two of which are illustrated in figure 47. The neuron shown in black has endings in the siphon which are sensitive to touch. At its other end, within the ganglion, the black neuron is connected by stimulatory synapses (curved lines in figure 47) with another neuron, shown in gray. The gray neuron, known as a motor neuron, has an axon extending out toward the gill muscles. When the siphon is touched, the black sensory neuron fires and sends a pulse down its axon to the synapses in the ganglion. This causes a release of the excitatory neurotransmitter called glutamate, which then triggers the gray motor neuron to fire, leading to withdrawal of the gill. The reflex response depends on this precise

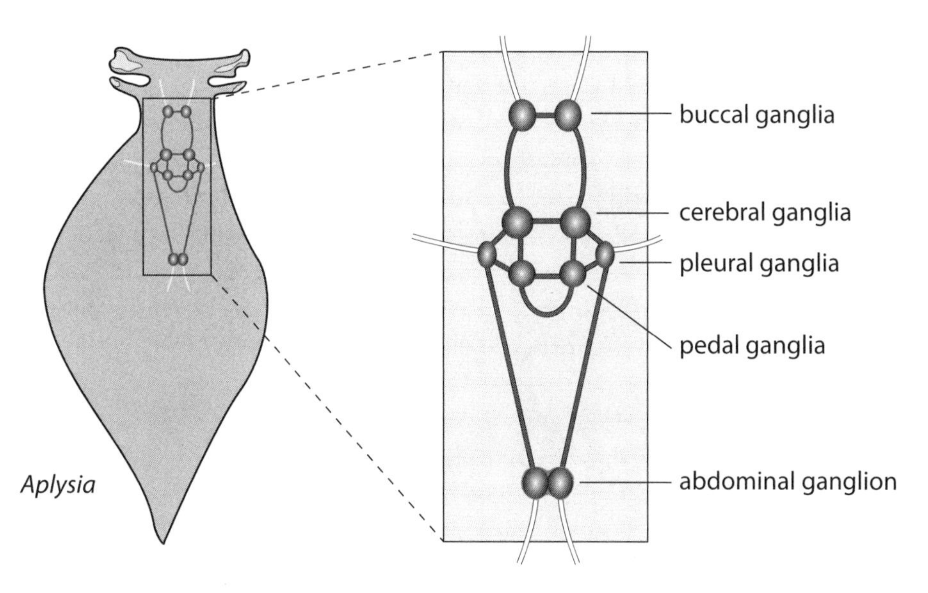

Figure 48. The brain of *Aplysia*.

set of neural connections and synapses, starting at the siphon, leading to the abdominal ganglion, and then on to the gill.

The gill-withdrawal reflex is just one of the many ways the sea slug can respond to changes in its environment. Figure 48 shows the organization of the slug's entire central nervous system. There are nine clusters or ganglia, each containing several thousand nerve cells. Different types of sensory input tend to travel to different ganglia. The gill-withdrawal reflex, for example, involves touch receptors in the skin sending their signals to the abdominal ganglion. Signals from the eye go to other ganglia, the cerebral ganglia, while those from the mouth travel to the buccal ganglia.

The organization of the slug's brain into different neural clusters or ganglia reflects the correlation of events in its environment. When an object touches the animal, it triggers nearby receptors in the skin. Other receptors of the animal are not affected. Thus, neighboring regions of the skin are most often triggered together: their activity tends to be highly correlated. This is mirrored in their pattern of connections—their axons converge on the same region of the abdominal ganglion where they are connected through synapses to nearby motor neurons. So, tightly associated inputs from sensory areas tend to be closely linked in the nervous system.

This makes sense because similar inputs often require similar responses and this is facilitated by the relevant neural termini being in proximity, allowing appropriate synapses to form.

The same principle applies to other sensory modalities. If an object enters the visual field, many eye neurons are triggered whereas other neurons remain silent. Similarly, if the animal eats something, a collection of mouth neurons are triggered while eye neurons are oblivious. There is a tighter correlation between neurons belonging to the same sensory system being triggered than between those from different ones. This pattern of correlation is reflected in the spatial organization of the animal's nervous system, with outputs from eye sensory neurons, for example, going to the cerebral ganglia, and those from the mouth going to the buccal ganglia.

The ganglia do not operate in isolation, however. What a slug eats may be related to what it sees—a tasty piece of seaweed looks different than an unsavory lump of rock. This is because there are characteristic associations between the way objects reflect light and what they are made of. Such relationships can be captured by the slug integrating information from the different ganglia: the buccal and cerebral ganglia in this case. The slug may learn, for example, that the appearance of seaweed is a good predictor of food in its mouth, whereas this does not apply to a rock (we will return to how such learning happens in the next chapter). There are therefore many axons that run between the ganglia, carrying information from one to another, and enabling responses to wider associations and distinctions. In this way, the anatomy of the animal captures relationships in its surroundings at a range of different scales.

All of these anatomical relationships are established in the animal through the process of development. The organization of the slug's nervous system, from the pattern of ganglia to the individual sensory and motor neurons with their initial sensitivities and connections, arises as it develops from an egg. The internal organization that develops then allows the slug to discriminate between different features in its environment and respond accordingly. Like a plant, the slug carves up the world by first carving up itself.

Patterns in Time

As well as responding to individual events, animal nervous systems can also respond to sequences of events over time. If you repeatedly stimulate the slug's siphon, the gill-withdrawal reflex progressively weakens, just as the mimosa leaves become habituated to continuous buffeting by wind. It is a case of what Marcel Proust called the anaesthetizing effect of habit—we downgrade our response as something we percieve as harmless becomes familiar.

In all forms of habituation the effect of the stimulus has to persist for some time so that it can weaken the next response. In the case of the slug, this persistence comes in two forms—short term and long term. Stimulating the siphon forty times consecutively results in habituation that lasts about one day. In other words, the slug is less responsive to siphon-touching for a day, but then its sensitivity returns to normal. However, if you stimulate the slug ten times every day over four days, the habituation lasts for weeks—the reduced responsiveness of the slug persists for much longer. What underlies these short- and long-term changes?

Short-term habituation entails changes in the strength of the synapses in the ganglion (figure 49). Repeated firing of the sensory neuron (black) leads to a weakening of synapses that connect it to the motor neuron (gray), and reduces the effectiveness of the signal. This is symbolized by the reduced size of the synaptic connections in the middle diagram of figure 49. The weakening of the synapses involves a molecular change in the terminals of the black neuron; following the arrival of an electrical pulse, they release less of the excitatory neurotransmitter glutamate. After a period of time, each synapse returns to its normal strength, accounting for the short-term duration of the change in sensitivity.

Long-term habituation results in a more lasting change in the synapses. Not only do the synapses become permanently weakened but the number of synapses drops—there is a change in anatomy as well as in the functioning of each synapse. This circumstance is shown in the diagram to the right side of figure 49; only one synapse now connects the black and gray neurons. These longer lasting changes in synapses involve genes being switched on or off in the nucleus of the black sensory neuron. Repeated stimulation of a synapse initially results in a temporary weakening, as with short-term habituation. This marks the synapse as being changed in some way. If the pattern of stimulation is repeated over several days, signaling molecules are repeatedly sent from the weakened synapse to the nucleus of the black sensory neuron. This leads to particular genes being switched on or off, which eventually leads to more permanent weakening of the marked synapse and sometimes its elimination. As with plant flowering, longer term persistence involves gene regulation.

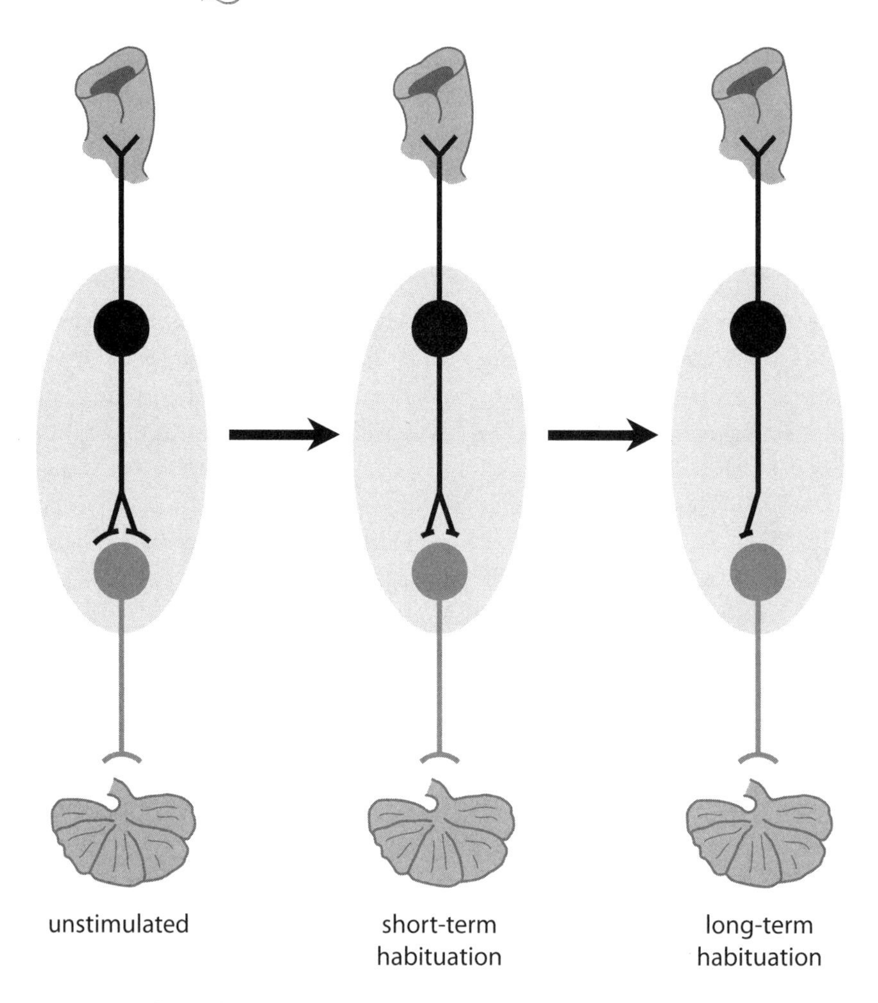

Figure 49. Effects of short-term and long-term habituation.

So far, we have looked at how the sensitivity of the slug can be reduced, but there are also situations that have the opposite effect. If instead of applying a relatively harmless stimulus, like touching the siphon, you do something very unpleasant to the slug, such as give its tail an electric shock, it goes into a heightened state of awareness. It now responds more dramatically to its surroundings, and when its siphon is next touched, it withdraws its gills in an exaggerated manner. The slug is said to have been sensitized by the earlier shock. Like habituation, sensitization involves persistent changes, but in this case the responsiveness increases, rather than diminishes.

The mechanism of sensitization is in many ways complementary to that of habituation. In short-term sensitization, selected synapses are temporarily strengthened, corresponding to an increased release of neurotransmitters when an electrical pulse arrives. In long-term sensitization, not only do the synapses strengthen more permanently, but more of them are created, increasing the number of contacts among the nerve cells. For example, instead of two synapses between the black and gray nerve cells, the number might increase to four or five. As with long-term habituation, this involves activation of genes through signaling between the synapses and the nucleus.

Human Responses

Though far more elaborate than the slug's head ganglia, the human brain shares some basic features of organization. As with the slug, the main function of the human brain is to integrate sensory information and generate appropriate responses. Consider our response to touch. Touch-sensitive neurons in our skin have axons that extend to the spinal cord, a long column of neural tissue housed within the vertebrae (fig. 50). For simple reflexes, the spinal cord acts as the meeting place for sensory and motor neurons, similar to the role played by the abdominal ganglion in the slug. In the knee reflex, hitting the knee triggers sensory neurons which send a signal along their axons to the spinal cord. There the axon termini are connected by synapses with motor neurons, which then relay the signal back to the knee, triggering muscle contraction and leg movement.

The brain only comes into play with more complex responses involving higher levels of integration. When you pick up an apple, you feel the size and shape of the fruit in your hand and this in turn influences your grasp. Without flow of information from your hand to your brain and back, your grasp may become too loose and you could drop the apple. This is far more complicated than a simple reflex response. When you grasp the apple, sensory signals from your hand arrive at the spinal cord and are relayed through more neurons up to the brain. There the signals are integrated in a range of different areas. One of these is a strip of the outer layer of the brain, or cortex, known as the primary somatosensory area (fig. 51, *top*).

The somatosensory area has a remarkable structure that was revealed by the neurosurgeon Wilder Penfield and colleagues, working at McGill

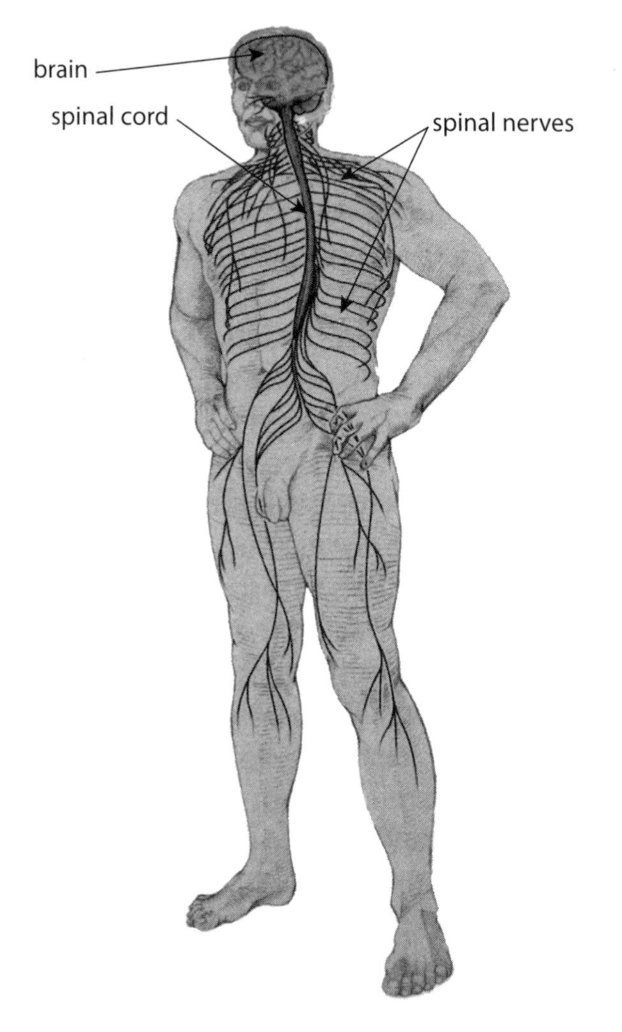

Figure 50. The human nervous system.

University in Montreal during the 1930s and 1940s. Penfield was operating on patients with brain tumors and wanted to delimit which areas of the brain were being affected, so that he could determine the likely impact of the tumor's removal. He found that when he electrically stimulated one region of the somatosensory area, the patient would report a tingling in the hand, while stimulating another region would create a sensation in the leg. The sizes of these regions in the brain were not in direct

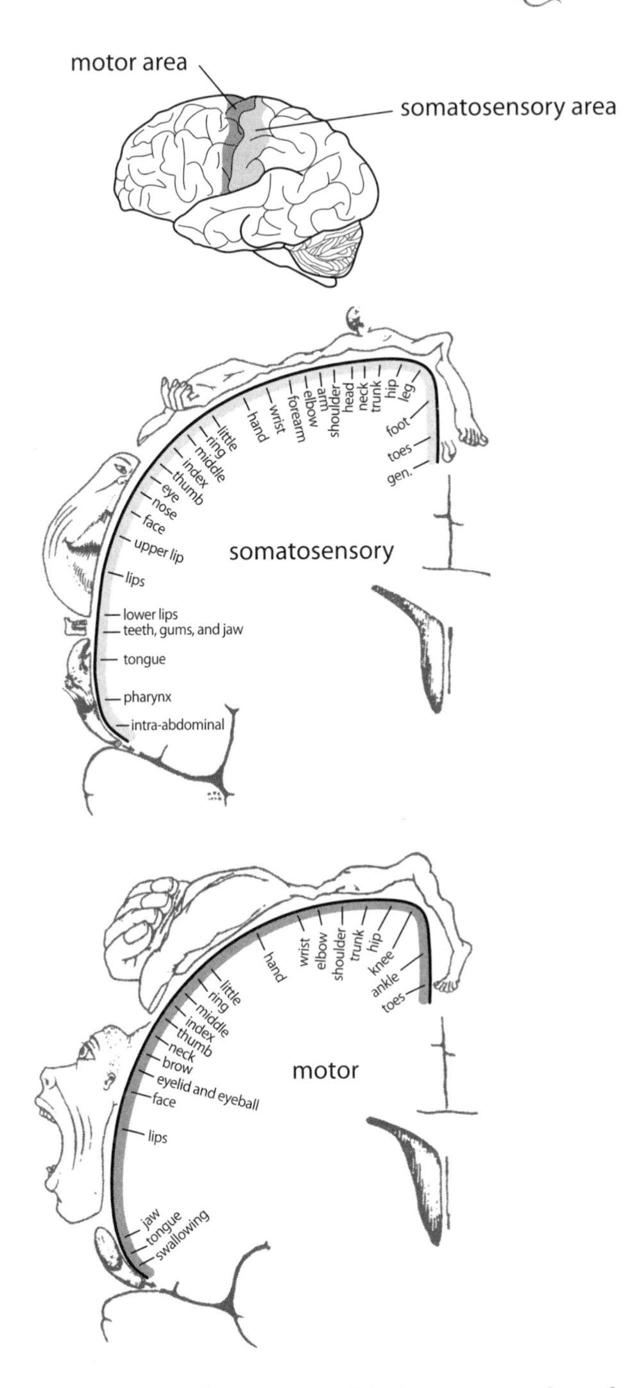

Figure 51. Somatosensory and motor areas of the human cortex shown from the side (*top*) and as cross-sections (*middle* and *bottom*).

proportion to the body parts; the region relating to the hand, for example, was much larger than that of the leg. Penfield illustrated this by drawing the body parts alongside the somatosensory area, enlarging or shrinking them according to the extent of the area occupied (fig. 51, *middle*). The result was a distorted homunculus with large hands and lips, and tiny legs. The distortion arises because the size of each region in the brain is related to number and complexity of its inputs; we have a much greater density of touch receptors in our hands than our legs.

Even though the somatosensory map is a distortion, the order of its regions largely mirrors that of the body. The region corresponding to the leg is next to that of the foot. Similarly, the region corresponding to fingers from the same hand are near each other. As for the slug, these spatial arrangements reflect correlations in the pattern of inputs—when we hold an apple, touch receptors throughout our hand are stimulated together while others are not affected. If we step into a bath, the receptors in our legs and feet are triggered together while others remain silent. Our neural organization reflects the relationships we most commonly encounter in our surroundings.

I have concentrated on touch, but the same principles apply to other sensory systems. We typically look at an apple as we grasp it. Light reflected from the apple triggers receptive cells in the retina (photoreceptors) and these signals are then relayed to the brain where they are integrated at various levels. The brain has specific areas that respond to these visual inputs. As with the somatosensory area, many of these visual areas are organized as maps that relate to their pattern of inputs: the various regions of the retina in this case.

The main purpose of integrating all of this sensory information is to respond with appropriate actions. At the sight of an apple we may extend our arm and grasp it with our hand. And we may then strengthen or weaken our grip according to the information we receive from our hand. All of these acts involve signals from the sensory areas of our brain influencing neurons that modify what we do.

Just as there are regions in the brain dedicated to sensory information, there are regions dealing with our actions. A good example is the primary motor area. Penfield found that when he stimulated this area, which lies just in front of the somatosensory area (fig. 51), particular joints would become active. A finger might twitch or an ankle might flex. This happened because signals from the stimulated motor area traveled through particular paths to enter the top of the spinal cord. They then went down the chord and triggered motor neurons that exited at particular levels, leading to specific muscle contractions. Again, Penfield found that the regions of the motor area were organized in a way that mirrors that of the body; the regions controlling the hand and fingers were close to one another, while those controlling the foot were next to the toes (fig. 51, *bottom*). This makes sense because our actions most commonly involve coordinated movement of nearby parts. When we tighten our grip on an apple, all of the fingers in our hand move together and this is reflected in the organization of the brain regions that drive these movements.

The human nervous system is organized at a whole series of scales, from the overall architecture of the nerves going to and from the spinal cord, to the brain with its various areas and subregions. Much of this anatomical complexity arises through the process of development. In chapter 4 (pages 107, 108), we saw how the embryo forms a neural tube that gradually becomes subdivided and elaborated to produce the spinal cord and brain. A human being is a walking tapestry of neural patterns that owes its origin to the process of elaboration that takes place in the womb.

Carving up the World

All of the responses we have covered in this chapter allow an organism to distinguish particular events, or sequences of events, in its environment. *Escherichia coli* can tell whether it is swimming toward or away from food, a plant can tell long from short days, a Venus fly trap can discriminate between an insect and a gust of wind, a sea slug can distinguish the touch of a predator from the harmless buffeting of the sea, and a human can tell whether there is an apple or piece of rock in their hand. These responses allow individuals to categorize the world, to carve it up according to particular events. Such categorizations can occur because the world is not homogeneous but highly organized at many scales. Matter collects to form suns, planets, rocks, seas, and living beings. The motion and interactions among the various collections of matter lead to a whole series of correlated changes. As our planet turns, the illumination from the sun at many points on Earth's surface fall and rise together, with entire countries being plunged into darkness or light. As an apple falls, all of the atoms within it move downward in unison. Our environment is a rich setting, involving many different types of correlated change.

Instinctive responses help an organism distinguish some of these correlations from others, and enable it to respond to those events that are most important for its survival and reproduction. A Venus fly trap snaps shut when the collective movement of the molecules in an insect triggers hairs on the leaf. Other changes around the plant, such as the movement of air molecules, do not trigger the hairs. The plant is dissociating one type of environmental circumstance (a moving insect) from another (moving air). It is able to do this because the forces exerted by the tightly associated molecules of an insect are stronger than those of a gas. And by responding to a sequence of such events rather than one, the Venus fly trap can further distinguish an insect from some sand that happens to be blown against it. The same ability to discriminate applies to animal responses. Touching a sea slug causes it to retract, while the water that normally circulates around it has no effect; collision with the tightly associated molecules in a solid has a greater impact than a liquid, and is more effective in triggering sensory nerve endings in the skin. By responding to a sequence of such events, the slug may further discriminate between what is likely to be harmless or not. Even the humblest of organisms categorizes the world in time and space.

But there is another aspect to such categorizations. The ability of individuals to respond and discriminate among events in their surroundings depends on their own internal distinctions or subdivisions. A plant's response to day length is woven into its anatomical arrangement—the organization of the plant as a series of leaves, stems, and growing buds. The response of a Venus fly trap depends on the spatial arrangement of the sensitive hairs and the structure of the leaf. The response of a sea slug or human depends on having specialized nerve cells with particular sensitivities and connections. These are organized at many different scales, reflecting the range of correlations in the environment that are encountered by the animal.

We have seen in previous chapters how this individual organization arises through development. The resulting internal arrangements allow each individual to respond in particular ways. But an organism's responses can also feed back to influence their own organization. Flower production may switch on in a weed in response to day length. The strength or number of a slug's neural connections may change in response to touch. In all these cases, the response leads to a persistent or lasting change in the organism. These organizational changes may take place while the organism is still growing and developing. A plant produces flowers as it grows and a slug is able to respond to its surroundings before it reaches adulthood. There is no sharp line between development and the responses it engenders. This is because internal arrangements are not fixed but to some extent can change in response to the surroundings, even while development is taking place. Our instinctive responses are not only enabled by development, they are woven into it.

All of the responses we have covered so far have been prescribed by the evolutionary past of the organism. Over many generations, particular responses have been favored by natural selection, such as the way a bacterium responds to sugar, a plant to day length, or an animal to touch. Yet many animals also exhibit responses to environmental elements never encountered before in their evolutionary past. A dog may learn to recognize its master's voice, a sound that its wolf ancestors were never exposed to. Or a rat may learn a path through a maze that would never have been experienced by its predecessors. These processes involve learning new relationships between elements of the surroundings, like the relationship between a sound and a person, or between a set of turns in a maze. Plants and microbes do not seem to have this ability to learn new relationships. Grass does not learn to duck down when the sound of a lawn mower starts. This is not just because of restrictions on sensing or responding to the sounds—it is because plants seem unable to learn new relationships within a lifetime. Even though plants and microbes can respond in very elaborate and sophisticated ways to their surroundings, their responses are always instinctive, based on experiences encountered in their evolutionary past.

Being able to learn new relationships is something peculiar to animals with elaborate nervous systems, like flies, slugs, dogs, and humans. So how does it come about? As we will see, it involves many of the components described in this chapter, but put together in a particular way. It arises through our third case of life's creative recipe.

SEVEN

The Neural Sibyl

Among the most striking paintings by Michelangelo are the five sibyls on the ceiling of the Sistine Chapel (fig. 52). Sibyls were divinely inspired women of antiquity who had the legendary power of being able to see into the future. In Michelangelo's depictions you see the sibyls holding special books in which prophecies were written. Sibyls were consulted much as fortune-tellers are today, to determine what lay ahead. We place a high value on being able to predict the future because if we know what is going to happen, we can prepare ourselves and adjust our actions accordingly.

Predictions about the future are not just an attribute of sibyls, prophets, and fortune-tellers; they are part of everyday life. We predict so often that we are hardly aware of it. When you turn a page of this book you expect that there will be more writing on the next. If you were to find blank pages you would be surprised and wonder whether there was a fault in book production. Similarly, if you put the book down somewhere and go for a cup of tea, you expect to find the book where you left it when you come back. It might seem rather grand to call such modest expectations predictions, but that is what they are. Every expectation we have about the future is a prediction, an anticipation of what will happen next. Without making continual predictions, our life would be impossible; we would be in a state of continual anxiety, never knowing what lay round the corner or what to expect.

Making predictions is not unique to humans; many other creatures have the same ability. If you are about to throw a ball to a dog, you can see

Figure 52. *Libyan Sibyl*, Michelangelo Buonarotti, 1511.

from its eyes and body posture how it anticipates or predicts when and how you are going to throw it. Or if a dog owner puts a coat on to go out, their dog may get very excited, anticipating that it is going for a walk. A dog continually forms expectations about what lies ahead, just as we do.

The ability of humans and other animals to predict is greatly influenced by learning. A newborn baby does not have the elaborate expectations of an adult because it still has to acquire them by learning from experience. Similarly, if you throw a ball to a young puppy, it does not look at you and anticipate as a more experienced dog does. Improving predictions through learning starts to come about immediately after birth—a newborn soon learns when to expect food, and further expectations rapidly follow. Some occurrences, however, are more predictable than others. We readily predict that the sun will rise tomorrow or that spring will arrive in a few months, but we may be much less certain about whether it will rain at three o'clock tomorrow afternoon or what will be the winning number of a lottery. Such knowledge might be very valuable to us but it defeats our predictive powers. The magic of the sibyls derives not from their ability to predict, for that is shared by all of us, but from having predictive powers that surpass what we can normally achieve.

Unlike sibyls, we and other animals acquire predictive abilities not with magical powers or by consulting prophetic tomes, but through the neural books of experience. How do these books get written? In the previous chapter we saw how our responses may change through the modification of neurons. A sea slug's sensitivity to touch may increase or decrease by the modification of synapses through earlier experiences. These are instinctive changes in our responses, highly constrained by the evolutionary past. But what about learning new things? Our ancestors of ten thousand years ago did not have books, and the wolflike ancestors of dogs did not have balls thrown to them. Yet humans and dogs acquire particular expectations for each. What is the neural basis of such learning? Trying to answer this question takes us to the heart of how we acquire knowledge about the world. As we have seen in previous chapters, it helps to consider some simplified examples when tackling larger questions. A good starting place for the neural foundations of learning is Pavlov's dog.

The Prophetic Dog

The Russian scientist Ivan Petrovich Pavlov began his studies on learning in about 1900, when he was in his early fifties. By that time he had already carried out pioneering work on the digestive system. During these earlier investigations, Pavlov had developed a simple procedure that allowed him to measure the amount of saliva produced by a live dog. When food was presented to the dog, Pavlov could readily determine how much more saliva it produced. Pavlov noticed that many factors could trigger the dog's salivation, as long as they were associated with presentation of food: "Even the vessel from which the food has been given is sufficient to evoke an alimentary reflex complete in all its details; and, further, the secretion may be provoked even by the sight of the person who brought the vessel, or by the sound of his footsteps."

Pavlov decided to study this systematically. He chose a stimulus that would not normally elicit greater salivation, such as the sound of a bell; Pavlov rang the bell every time the dog was about to be given food. The dog eventually came to salivate more whenever the bell was rung, even if no food followed. Its response had become conditioned by the bell. Such Pavlovian conditioning is often interpreted as showing that a dog learns the association between the bell ring and the food. Indeed, Pavlovian conditioning is frequently referred to as being an example of associative learning. But there is another possible interpretation. Perhaps the dog is learning to predict, rather than learning an association. From the sound of the bell it is possible to predict that food is on its way. Maybe that is why the bell comes to elicit greater salivation. What is the difference between association and prediction?

After the wheel stops spinning in a game of roulette and the ball falls into a pocket, the croupier announces the number. We may say that there is an association between the number the croupier calls and the number of the pocket that the ball falls into. The two go together. But we would hardly say that the croupier predicts which pocket the ball will land in. This is because the croupier gives the result after the ball has settled. If the croupier correctly called the number before the ball stopped, then we would say that the croupier had predicted the pocket correctly. Hence, our notion of what is or is not a prediction is all about timing. By contrast, we can discuss associations with no reference to temporal order. For example, we may note an association between the croupier's call and the ball's landing pocket without regard to whether the ball settles before or after the croupier makes the announcement.

If Pavlov's dog is learning an association then it should not make a difference whether the bell is rung before or after the food is presented. However, it turns out that it does. If the bell is always rung just after, instead of before, the food is offered, the bell does not come to elicit greater salivation when sounded alone. Indeed, the bell ring instead tends to have an inhibitory effect on salivation. It seems that the predictive value of the bell is what matters, rather than just the association of the bell with the food. If the bell is rung before the food, it signifies that food will appear; the dog then salivates more in preparation for this event. If the bell is rung after the food has been offered, it signifies that the food will disappear as the dog eats it up, so the dog prepares for this event by salivating less. The dog is learning predictions, not just associations.

The idea that the dog's response involves prediction is consistent with another type of experiment. This experiment was carried out by Leon Kamin working in Princeton in the 1960s with rats, although I will continue discussing dogs in order to illustrate his results. Suppose that instead of a bell ring, we condition our dog with another type of stimulus, such as touching the paw. If we always touch the paw before food is given, the dog eventually learns to salivate more after the paw is touched—this is the traditional Pavlovian response you would expect. We can continue with conditioning like this until the dog is fully trained to respond to the paw touch. What happens if we now repeatedly apply two stimuli simultaneously, both touching the paw and ringing the bell, just before food is given? If the dog is learning by independently associating each signal with the food, then it should learn to respond to the bell as well as touch, because both are now occurring together before the food. But the dog does not learn to salivate in response to the bell alone, though it still salivates in response to touch alone. This makes sense if the dog is learning according to the predictive value of the stimulus. Having learned that the paw touch signifies that food is on its way, the bell now adds no new information; it has no predictive value, so the dog does not learn to connect it with food. Again, timing is important. If instead of ringing the bell at the same time as touching the paw, we ring the bell earlier, then the dog learns to salivate in response to the bell, even if it has previously been trained with the paw touch. The bell is now providing additional information—it is an earlier indication that food is coming, so the dog learns this new relationship. Pavlovian conditioning is all about prediction.

It seems that the dog's brain is making prophecies based on its past, learning from prior experiences to anticipate the future. What sort of mechanism might underlie such learning? While Pavlov was carrying out his studies in Russia, other scientists like Santiago Ramón y Cajal in Spain and Charles Sherrington in England were looking into the details of how neurons work. They showed that a key feature of neurons lies in the way they trigger each other through synapses. How do these investigations into neurons relate to Pavlov's results? Pioneering attempts to answer this question were made in the late 1940s by two scientists: a Polish neurophysiologist Jerzy Konorski, and a Canadian psychologist Donald Hebb. In his book *Conditioned Reflexes and Neuron Organisation* (1948), Konorski showed how Pavlov's results could be interpreted as involving specific changes in the formation and number of synaptic connections between neurons. Donald Hebb independently arrived at a similar conception in his book *The Organisation of Behaviour* (1949). Yet in spite of these pioneering studies, it remained unclear how synaptic changes were brought about and how they might lead to improved predictions. It wasn't until forty years later that the story of conditioning could be taken further, when neurons involved in predictive learning were identified.

Predictive Neurons

In the mid-1980s, Ranulfo Romo and Wolfram Schultz at the University of Fribourg, Switzerland, were recording electrical signals from particular neurons in a monkey's brain as it learned tasks. These neurons are located in a region known as the midbrain, and they produce a neurotransmitter called dopamine at their terminals. Recall that neurotransmitters are substances released from axon terminals that travel across synaptic gaps to influence neural activity on the other side.

The firing of the dopamine-releasing neurons that Romo and Schultz were studying seemed related to the monkey receiving rewards. A box was connected to the monkey's cage that sometimes contained a reward, such as a piece of apple. The monkey could not see the contents of the box and so initially had no idea whether it contained food. But if the monkey reached into the box, it would sometimes feel a piece of apple and retrieve it. Romo and Schultz found that the neurons they were recording from fired at a much higher rate when the monkey reached in and found the piece of apple (fig. 53, *top panel*). At first they suspected the firing might be related to the arm movements of the monkey. But they then showed that the neurons didn't increase their firing rate when the monkey

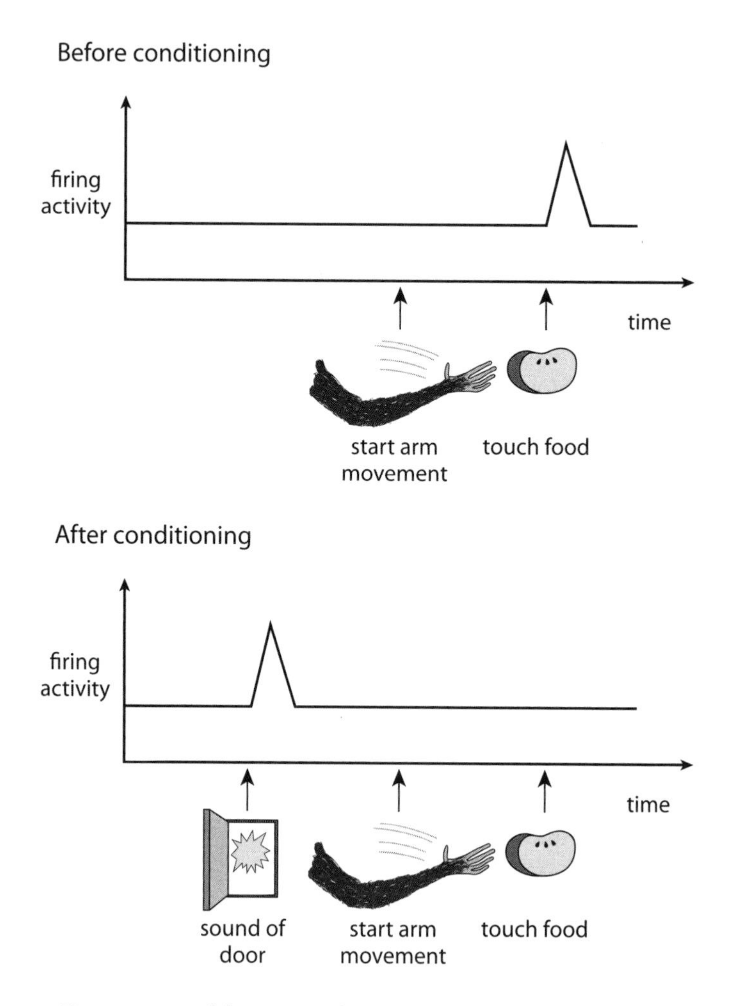

Figure 53. Firing activity of dopamine-releasing neuron before and after conditioning.

did not find a piece of apple in the box. It appeared that firing of the dopamine-releasing neurons was linked to touching the apple, and not to the arm movements.

They then changed the experimental arrangement in order to condition the monkey with the sound of an opening door. The box now always contained apple but it was closed off from the monkey by a small door. The monkey soon learned that when it heard the sound of the door opening, it would be able to reach into the box and get a piece of apple. Romo and Schultz then made a remarkable observation. They found that neurons that were previously triggered by touching the apple were now triggered by the sound of the door opening (fig. 53, *bottom panel*). The neurons had shifted their response back in time; they were triggered by the predictive signal, the sound of the door, rather than the apple itself.

To understand this, it helps to rephrase the monkey experiment in terms of three notions—*rewards*, *expectations*, and *discrepancies*. Initially, the monkey has no expectation of a reward (the piece of apple) when it hears the door sound. Consequently, when it happens to find the door open and is able to reach in to get a piece of apple, there is a discrepancy between what the monkey expects (door closed and no reward) and what it finds (door open and reward). The effect of learning is to eliminate this discrepancy—the sound of the door leads the monkey to expect that it will get a reward. However, the monkey still does not know when to expect the door sound. It has no way of knowing when the door sound is going to be made, so there is still an element of surprise, or a discrepancy in expectations, at that point. The discrepancy has effectively been shifted from the time when the monkey gets the reward to an earlier time, when the door sounds. The neuronal activity that Romo and Schultz were measuring seems to be related to the timing of discrepancies, to when the monkey is surprised. The problem is to explain how such a shift in timing could arise through neural interactions.

Learning from Discrepancies

An ingenious solution came a few years later from computational neuroscientists Read Montague, Peter Dayan, and Terry Sejnowski working in the United States. Their solution integrates the experimental findings of Romo and Schultz with a mathematical theory of learning, called Temporal Difference learning. TD-learning had been formulated by Richard Sutton and Andrew Barto several years earlier. It is worth going into the mechanism proposed by Montague, Dayan, and Sejnowski in some detail because it serves to highlight the key principles involved in learning.

A simplified version of their scheme is illustrated in figure 54. You can see two neurons that receive various inputs. I have called the upper one an *expectation neuron*. By giving it this name I do not mean to imply that it has special psychological properties that allow it to expect one thing or

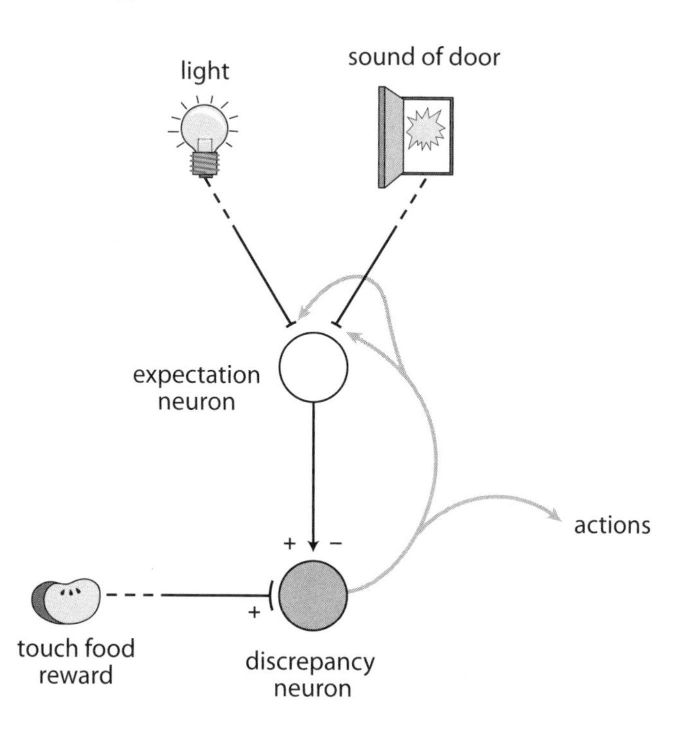

Figure 54. TD-learning scheme showing neural connection strengths before conditioning.

another. It is a neuron like any other, with inputs and outputs. But as we will see, the role this neuron plays in the scheme means that it reflects the expectations of the monkey. The expectation neuron receives inputs from a variety of sensory signals, such as the sound of the opening door, or visual stimuli, such as a light flashing. These signals are connected to the expectation neuron by weak synapses, shown with small knobs at the axon ends. Initially these signals do not trigger the expectation neuron very much, if at all.

The lower neuron, shown in gray, is called a *discrepancy neuron*. Again, it is a neuron like any other and its name only reflects the role it plays in our scheme. The discrepancy neuron receives a stimulatory input (indicated by a positive sign) from a neuron sensing a reward, as when the monkey touches the apple. The reward signal is connected to the discrepancy neuron with a strong synapse, so touching the apple greatly stimulates firing of the discrepancy neuron. The discrepancy neuron also receives input from the expectation neuron. This input can either be stimulatory or inhibitory, as indicated by the positive and negative signs. Whether the input is stimulatory or inhibitory depends on whether the firing rate of the expectation neuron is increasing or dropping. If the firing rate of the expectation neuron is rising, then it stimulates the discrepancy neuron. But if the firing rate of the expectation neuron is falling, then it inhibits firing of the discrepancy neuron. The situation is rather similar to what we encountered with the swimming bacterium *Escherichia coli* in the previous chapter. Recall that it changed its swimming behavior according to whether the level of sugar in its surroundings was increasing or decreasing. Like the sugar concentrations, it is the relative change in the firing of the expectation neuron that matters. A key point to remember is that if firing of the expectation neuron is rising, then the discrepancy neuron is stimulated, while if it is falling then the discrepancy neuron is inhibited.

The final element in the scheme is that the firing of the discrepancy neuron feeds back to influence the inputs to the expectation neuron. It does this in a rather special way. Strong firing of the discrepancy neuron tends to strengthen synapses from neurons that had been firing just before the discrepancy neuron fired. If the neuron sensitive to the door sound had been firing prior to the discrepancy neuron's activity, then the strength of its synapse increases slightly when the discrepancy neuron fires. However, if the door-sound neuron had not been firing prior to the discrepancy neuron firing, then its synapses with the expectation neuron do not increase and may even decrease in strength.

We are now in a position to see how predictive learning can work. Suppose the monkey touches the apple after hearing the door. Feeling the apple makes the discrepancy neuron fire at a high level (because of the strong stimulatory synapse). Any input to the expectation neuron that fired just before this will therefore have its synapse with the expectation neuron strengthened slightly. This situation applies to the door-sound neuron but not to the light-flash neuron. Even though both the door-sound and light-flash neurons can provide inputs to the expectation neuron, only the synapse with the door-sound neuron is strengthened because it fired just before the discrepancy neuron (which was in turn triggered by reward delivery). Whether a synapse is strengthened depends on whether it was being triggered just before reward delivery.

Having recently fired, the door-sound neuron is effectively boosting its own strength with the help of the discrepancy neuron. This self-boosting continues with further rounds of door opening and food retrieval, leading to a stronger and stronger synapse forming between the door-sound and expectation neuron as indicated by the longer length of the synapse in figure 55. You might expect this self-reinforcement to continue indefinitely, with the synapse between the door-sound neuron and expectation neuron becoming ever stronger. This does not happen, however, because the door-sound neuron eventually becomes a victim of its own success, as we will now see.

As the strength of the synapse with the door-sound neuron increases, the firing pattern of the discrepancy neuron starts to change in two ways. First, the discrepancy neuron now begins to fire at a higher level when the door opens. This is because upon opening the door, the door-sound neuron fires and triggers the expectation neuron through its stronger synapse. The rise in firing of the expectation neuron in turn triggers the discrepancy neuron (recall that the expectation neuron has a stimulatory effect when its firing level increases). So the discrepancy neuron now fires at a higher level when the monkey hears the door open.

Secondly, the firing pattern of the discrepancy neuron changes at the time of reward delivery. As before, touching the apple stimulates firing of the discrepancy neuron. But as this is happening, there also is a fall in the firing level of the expectation neuron. This is because the monkey is no longer hearing the sound of the door opening. According to the rule described earlier, this drop in firing activity of the expectation neuron leads to inhibition of the discrepancy neuron. The discrepancy neuron now receives two inputs antagonistic to one another—a stimulatory input from the apple reward and an inhibitory input from the reduced firing of the expectation neuron. This inhibitory effect of the expectation neuron means that the discrepancy neuron fires less than it did the first time the reward was delivered. This in turn means less strengthening of the synapse of the door-sound neuron, counteracting the tendency for this neuron to boost itself.

We have a self-limiting system. With each experience, the firing of the expectation neuron tends to rise as soon as the monkey hears the door because of the synapse that has been reinforced. But a consequence of this early rise is a fall in the firing of the expectation neuron later on. It is an example of what goes up must later come down. The later drop in expectation neuron firing inhibits the discrepancy neuron, reducing the degree to which it promotes synaptic strengthening. The process continues in this

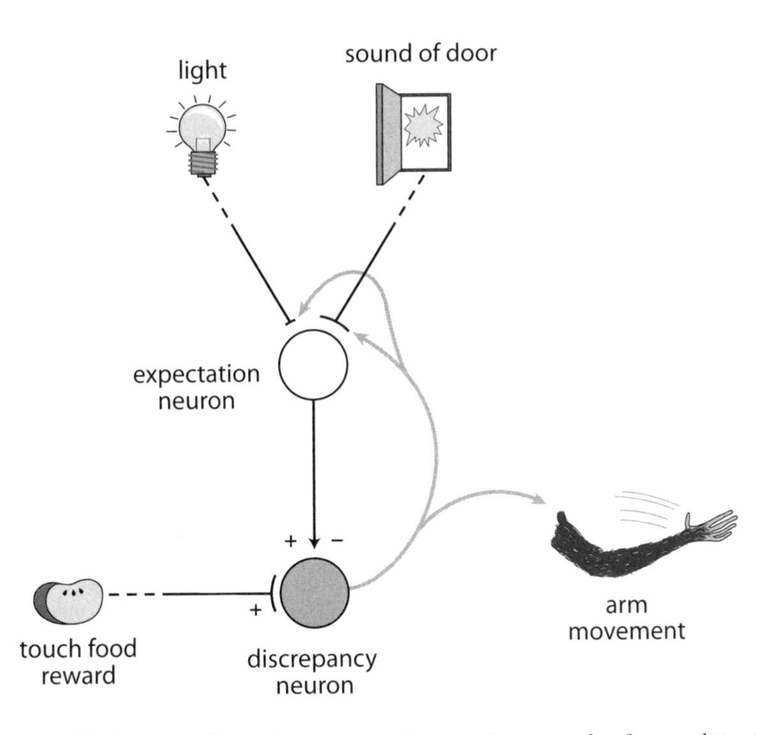

Figure 55. TD-learning scheme showing neural connection strengths after conditioning. Compared with figure 54, the synapse with door-sound neuron is stronger, and movement of the monkey's arm has now become coupled to firing of discrepancy neuron.

way until the inhibitory effect of the declining firing of the expectation neuron exactly counterbalances the stimulatory effect of the reward. At this point, firing of the discrepancy neuron is no longer raised when the apple is touched, so there are no further changes in synaptic strengths. The system has stabilized because the change in firing of the expectation neuron is now matched against the stimulation of the reward. The only time the discrepancy neuron now fires at a high level is when there is a rise in firing of the expectation neuron, caused by the monkey hearing the door sound.

This TD-learning scheme explains how the shift in firing time of the dopamine-releasing neurons studied by Romo and Schultz may arise. These neurons correspond to the discrepancy neuron in our scheme. Initially, the discrepancy neuron fires at a high level when the reward is sensed; but after learning, the raised firing is shifted back in time to when the monkey hears the sound of the door opening. This backward shift is exactly what is observed for dopamine-releasing neurons, suggesting that they are behaving like the discrepancy neurons in our scheme. The implication is that the release of dopamine from their terminals must somehow increase the strength of synapses. Exactly how this happens is not known. It may be that the dopamine release acts directly on the synaptic inputs or it may act more indirectly. Either way, what matters is that firing of these neurons somehow leads to a modification of synaptic strengths.

What good is this system for the monkey? For learning to be useful, it has to be linked to what the monkey does. This can be achieved through a connection between the output of the discrepancy neuron and the monkey's arm movements, shown by the arrow in figure 55. Initially, the monkey's arm movements are not linked to the sound of the door opening. But after conditioning, the monkey moves its arm when it hears the door sound. According to the TD-learning scheme, this happens because the raised firing of the discrepancy neuron comes to trigger actions that are likely to increase the chance of getting a reward. This process may occur through synaptic strengthening similar to that described for the expectation neuron, except that it now involves neural connections to particular actions, rather than external signals. (We will return to how actions may be reinforced in the next chapter.) The net result is that the triggering of the discrepancy neuron by the door sound triggers movement of the monkey's arm to retrieve the food.

Before finishing this account of TD-learning, I should clarify something I have glossed over. I assumed that firing of the expectation neuron falls at the same time that the reward is delivered. This ensures that the inhibitory effect of the expectation neuron counterbalances the stimulatory effect of the reward. But why should firing of the expectation neuron fall exactly when reward is delivered? According to the proponents of TD-learning, this is because the brain has a way of discerning the timing or duration of signals, not just their levels. For example, we might imagine that the door sound sets off a series of firing patterns of various durations in the brain, some lasting a short time, others a longer time. It would be as if the door sound reverberates for a range of times in the brain, providing a range of inputs to the expectation neuron. Suppose we could arrange that only the input signal that lasts until just before the reward is delivered gets its synapse reinforced; input signals with longer or shorter durations would not get strengthened by the discrepancy neuron. Such a system would ensure that firing of the expectation neuron falls at the same time that the reward signal arrives, because it is only inputs with this property that get strengthened.

Timing mechanisms like this may sound rather contrived, but they account for a key neural observation. Suppose that after conditioning, the reward is withheld from the monkey. The monkey reaches into the box but is disappointed to find no apple. Schultz and colleagues found that in this situation the firing rate of dopamine-releasing neurons falls at precisely the time that the monkey would normally have touched the apple. In other words, firing is inhibited exactly when the monkey was expecting to receive the reward. This result is what you would expect from the TD-learning scheme, because the firing of the expectation neuron is falling at exactly this time. This drop leads to inhibition of the discrepancy neuron which is now not counterbalanced by a reward signal (there is no apple), so the firing rate of the discrepancy neuron falls below its normal level. Reduced firing of the discrepancy neuron corresponds to disappointment of the monkey, just as raised firing corresponds to a pleasant surprise.

Pavlov and Punishments

TD-learning provides a general mechanism for predictive learning, involving a neural interplay between expectations, rewards, and discrepancies. The same scheme can account for the behavior of Pavlov's dog. Initially food appears out of the blue, without the dog expecting it. This discrepancy between expectation and reward reinforces synaptic connections from signals that precede it, like the bell sound. Through this process, firing of the dog's expectation neurons is raised when it hears the bell, leading to a shift in discrepancy neuron activity from the time of food delivery to the bell ring. The main difference from the monkey example is that instead of arm movements, the action triggered by the discrepancy neuron is increased salivation. Actually, Pavlov's dog is simpler than the monkey case because the link between discrepancy and action does not have to be learned—the salivation response was already in place before conditioning. Cases like Pavlov's dog where the action is already in place are called classical conditioning. By contrast, cases like the monkey retrieving the apple, where the action is also learned, are called operant, or instrumental, conditioning. TD-learning can account for both kinds of conditioning.

TD-learning can also explain why conditioning with one predictive signal, such as a paw touch, can block later conditioning with another signal, such as the bell sound. After a predictive signal like a paw touch has been learned, there is no longer a discrepancy at the time of reward delivery (the discrepancy neuron is not triggered at this time). There will therefore be no strengthening of synapses for signals that occur with the paw touch. The system has already learned what to expect and does not strengthen connections with inputs that have no further predictive value.

So far, I have described examples involving rewards like food. But exactly the same scheme can be applied to punishments, the opposite of rewards. Suppose that following a light flash, the monkey gets an electric shock rather than a piece of apple if it should put its hand into the box. The monkey soon learns not to put its hand in the box after a light flash. We can apply the same TD-learning scheme as before to achieve this result, but with a different type of discrepancy neuron. Instead of inputs from rewards like an apple, this discrepancy neuron would receive inputs from aversive signals like the electric shock. A further difference is that the action resulting from firing of the discrepancy neuron would involve avoidance rather than engagement. Learning would proceed as before except that actions immediately preceding firing of the discrepancy neuron would be inhibited. So, if the light flashes, the monkey would learn to avoid putting its hand in the box. Although such a scheme works in principle, the neurons involved have yet to be clearly identified. We therefore know less about the neural details of punishment as compared to reward learning.

TD-learning provides our best current model for how predictive learning can occur through neural interactions. I now want to use this example to look at the basic principles involved in learning. In what follows, I use terms also employed when describing evolution and development in order to highlight common fundamental principles, while recognizing that these processes differ in many respects as well.

Core Principles

Major changes in synaptic strength do not take place all at once during TD-learning; they depend on the cumulative effect of multiple experiences. We are dealing with a population of events, or many instances of door opening and reward delivery. As we have seen before, a population always assumes a context, like the turns of the roulette wheel or molecules in a tea cup. In the case of TD-learning, the context is the set of neural connections in the brain and the range of experiences encountered. Each experience on its own may cause only a slight change in synaptic strengths, but through multiple experiences a significant change occurs overall. If it were not so—if it just took one event to set synaptic strengths—then an animal would form its neural connections only according to its latest experience. It would rewire itself according to what happened most recently and erase preceding experiences. But our most recent experience is not necessarily the best predictor of what will happen next—winning once at roulette does not mean that we will win the next time. There are some cases, particularly those involving strong punishments, when a single experience may be enough for us to learn from—as expressed by "once bitten twice shy." But in our complex and unpredictable world it is usually better to learn on the basis of overall trends rather than according to the immediate past. The more uncertain we are of our surroundings, either because of a lack of experience or because of variability in our environment, the more it pays to learn in this manner. This form of learning is precisely what TD-learning achieves. By adjusting synaptic strengths a small amount with each event, it allows predictive trends, rather than a single event, to be learned. TD-learning is built on the principle of population variation.

TD-learning also depends on persistence. The system would not work if after every time the strength of a synapse was modified, it returned to its initial condition. As we saw in the previous chapter, changes in synaptic strength may persist for a short or long period. The first sequence of experiences during learning may lead to short-term changes in synaptic function, rendering a synapse either more or less effective. These short-term changes may then build up through further experiences to result in longer term anatomical changes, such as changes in synapse number. Longer term synaptic changes can last for many years, as illustrated by the following story told by Charles Darwin.

In 1836, after having just returned from his five-year voyage on the *Beagle,* Darwin decided to test the memory of his dog. He went to the stable where the dog was kept, and called out to him. The dog rushed forth and happily set off with Darwin for a walk, showing the same emotion as if his master had been away for half an hour. The sound of Darwin's voice still triggered the dog's response after a five year absence. Perhaps this story also tells us something about Darwin's brain. Darwin had been taught the classics at school, often learning forty or fifty lines from Homer before morning chapel. In Homer's classic story, *The Odyssey*, Odysseus returns to his home town after a ten year sea voyage. Odysseus dresses himself as a beggar to avoid being recognized, but this does not fool his old dog Argus who instantly pricks up his ears and wags his tail at the sound of his master's voice. Darwin thought that reciting Homer at school was a complete waste of time because he forgot every verse after two days. But perhaps the story of Odysseus established some longer term neural connections in Darwin's brain, and was unconsciously triggered when he returned from his long sea voyage, leading him to test the dog's memory. Darwin and his dog were perhaps more similar than even he realized.

Population variation and persistence are both key ingredients of learning. Without variation there would be no changes in synaptic strengths, and without persistence every synaptic change would vanish as soon as it was formed.

Reinforcement also plays a vital role in TD-learning. Firing of the door-sound neuron boosts the strength of its own synapse, with the help of the discrepancy neuron. Such reinforcement is essential for the system to operate; it is through this process that the door-sound neuron becomes more effective in triggering the monkey's response than other potential input signals like the light flash. Similarly, with Pavlov's dog, the firing of a bell-ring neuron tends to strengthen its own synapses. We can summarize such reinforcement with a positive feedback loop as shown to the left of figure 56, where the firing of a neuron promotes its own synaptic strengths.

Left to its own devices, reinforcement would lead to certain synapses growing stronger and stronger as experiences build up. The system would boost itself to ever higher levels of synaptic activity, resulting in a brain that would eventually buzz itself into a frenzy. This does not happen because the self-boosting eventually becomes a victim of its success. As the synapse with the door-sound neuron increases in strength, firing of the expectation neuron is also raised at the sound of the door. Consequently, there is a greater fall in the firing of the expectation neuron at the time of reward delivery, damping down the discrepancy neuron's activity and thereby reducing any further strengthening. Reinforcement has led to its

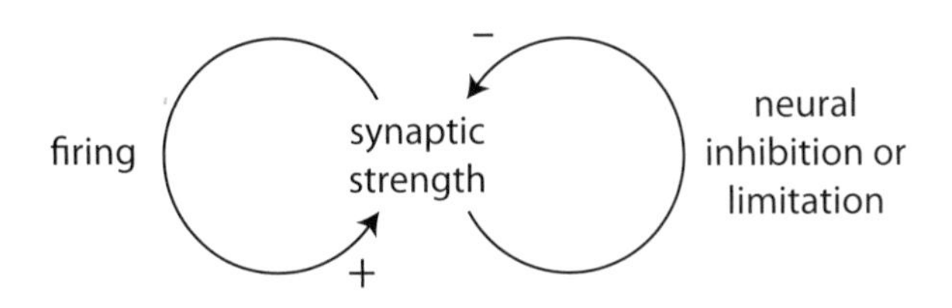

Figure 56. Interplay between reinforcement (positive loop) and competition (negative loop) for learning.

own inhibition or limitation, as shown by the negative loop to the right of figure 56. This is our familiar double feedback loop with reinforcement promoting competition (compare with fig. 12 on page 67); with competition in this case involving the inhibitory action of neurons. The overall effect of this combination of positive and negative loops is that expectations can be matched to the levels of reward or punishment.

We have arrived at the same core principles as those for evolution and development. At the heart of learning we have a double feedback loop of reinforcement and competition, fueled by a balance of variation and persistence. The difference is that in this case we are dealing with a population of experiences and neural interactions, instead of a population of individuals as in evolution, or a population of molecules and cells in development. The principles are similar, but the guise is different.

Of course there are many differences between evolution, development, and learning. Varying experiences are not the same as genetic variations in a population or collisions among molecules. And synaptic strengthening is very different from reproductive success. In spite of these numerous differences, there are nevertheless some overarching similarities in form—a similar set of basic feedback loops and ingredients.

It is no accident that we have arrived at some common principles. After all, the approach taken in this book has been to look at what various transformations might share. The aim of this exercise, though, has not been to look for similarities for the sake of it; but to help us comprehend the essence of living transformations. Traditional accounts of TD-learning do not present it as a double feedback loop between reinforcement and competition, fueled by a balance of variation and persistence. But by viewing it from this perspective, we more clearly appreciate both TD-learning's fundamental logic and its relationship to other processes.

I have presented TD-learning as an example to illustrate how common principles apply to a particular learning mechanism. But the same could be said for most mechanisms that have been proposed for learning. In a variable and uncertain environment we need to learn trends from a population of events rather than single instances. For learning to have lasting effects, there has to be some form of persistence. For some neural connections to boost themselves more than others, there has to be reinforcement. And reinforcement must limit itself through competition so that a match can be established between our experiences and what is learned. The details may vary from one learning mechanism to another, but the same core principles are always there.

A Neural Journey

It takes more than one neuron to achieve learning. In the case of TD-learning, interactions exist between the expectation and discrepancy neurons, as well as neurons sensing stimuli like touch, light, and sound, and neurons involved in actions like arm movements or salivation. These neurons may also assist each other's actions. The strengthening of the synapse between the door-sound neuron and expectation neuron depends not only on the firing of the door-sound neuron but also on the firing rate of the discrepancy neuron. The two firings together are necessary for synaptic strengthening. For this to occur, the outputs of the door-sound and discrepancy neurons must converge on the expectation neuron. As we have seen before, cooperation often involves physical proximity; in this case, particular neural termini and dendrites are brought together. Without this kind of proximity, each neuron would function in isolation and learning would be impossible. The principle of cooperation is a further key ingredient of learning.

Bringing together multiple components ushers in the possibility of numerous combinations. The human brain contains about one hundred billion neurons. Each neuron may have thousands of synapses, so the total number of synaptic connections is about a thousand times greater: one hundred trillion. This creates a vast space of neural possibilities. We have encountered large spaces before. When describing evolution we came across genetic space, the range of possible DNA sequences. For development we encountered developmental space, the range of possible cell states and their arrangements in an embryo. These are all vast hyperspaces with numerous dimensions. Similarly, we now have neural space, the enormous set of possible connections and firing states of neurons in the brain. This is the principle of combinatorial richness applied to neural connections.

We previously thought of evolution as populations journeying through genetic space, and development as an embryo traveling through developmental space. Similarly we may think of learning as a brain journeying through neural space. We are each born with a highly structured brain, corresponding to a particular location in our imaginary neural space. This location is slightly different for each of us, largely because of differences in genetic makeup that influence brain development in the womb. Following birth, we begin to interact strongly with our environment; each interaction shifts our position in neural space, taking each of us on a highly intricate, neural journey. Learning plays a major part in moving us along our journey by changing the strength and number of neural connections. The examples of Pavlov's dog and the monkey reaching for its reward illustrate how connections may change for a few interacting neurons. But the same sort of changes take place in parallel for numerous inputs, outputs, and connections in the brain, leading to an elaborate neural journey.

One of the main consequences of this neural journey is the improvement of our ability to predict what is likely to happen. We continually modify our expectations based on our history of experiences and are then able to cope better with our surroundings. But what keeps us moving in neural space? What provides the continual driving force for learning?

Staying on the Move

One of the key features of TD-learning is that it does not eliminate discrepancies; it shifts them. For the monkey experiment, the overall result is that the door sound, instead of touching the apple, comes to trigger the discrepancy neuron, so the discrepancy is shifted to an earlier time. Whereas the monkey previously got excited when touching the apple, it now gets excited at the sound of the door. According to the proponents of TD-learning, this is because one consequence of the discrepancy neuron firing is to excite or give pleasure the monkey. Indeed, the neurotransmitter dopamine released by the discrepancy neuron (as studied by Romo and Schultz) is thought to play a major role in drug addiction. Drugs such as cocaine and amphetamine are thought to exert their effects by enhancing dopamine action. Initially, touching the apple triggers the discrepancy neuron and dopamine release, and thus excites the monkey. After learning, the door sound is behaving like the apple was at the beginning; it triggers the discrepancy neuron and dopamine release. From this neural perspective, the door sound has become a substitute reward—it triggers the discrepancy neuron and dopamine release just as the apple did. If a new stimulus, like a light flash, occurs before the door sound, then the monkey would treat this situation just as if the light flash was a predictor of reward. But in this case the "reward" is not touching the apple but substitute reward of the door sound. If the monkey has many experiences of the light flash preceding the door sound, TD-learning ensures that the discrepancy neuron starts to fire even earlier, at the time of the light flash. We are building one expectation on another. Having learned to expect the apple reward based on the door sound, the neural system automatically learns to respond to factors that might allow it to predict the door sound. The drive for learning does not stop; it has instead shifted to another stimulus.

The ability of learning to create substitute rewards in this way is so prevalent that it can be hard to identify what constitutes an instinctive versus a learned reward. In many of Pavlov's experiments, the smell or sight of meat was used to condition other responses. You might imagine that dogs have an inborn salivation response to the presence of meat and learn signals, such as the bell ring, that act as predictors of this reward. But the response to meat turns out not to be inborn; it is a result of conditioning. If a dog is fed for a long time on milk alone after birth, then it does not salivate at its first smell or sight of meat. The smell or sight of meat start to evoke the salivary reaction only after the dog has been fed meat a few times. The odor and appearance of meat then act as predictors of rewards like a satisfying meal, and start to become rewards in their own right. What we normally think of as an instinctive or unlearned reaction, like salivation in response to meat, actually arises through conditioning. This does not mean that the dog is born without any instinctive responses. Inborn ways of responding to stimuli that act as rewards or punishments are determined by the neural connections established during development of the dog as an embryo. But as soon as a dog is born and exposed to its surroundings, it starts to build on these reactions through learning, establishing further sets of stimuli that act as rewards or punishments.

The phenomenon of one learned response forming the foundations for another is called secondary conditioning. It arises naturally from TD-learning because this mechanism operates in a relational manner. As discrepancies are encountered, expectations change and shift the discrepancies, introducing substitute rewards for the next round of learning.

Humans with their elaborate brains are expert at this game of adjusting expectations and rewards. We often treat money as a reward, working so as to get paid at the end of the month. Yet money only has value because of the things it allows us to buy, such as food or other goods. We may, for example, purchase an apple in a shop. Yet we don't usually eat the apple straight away; the reward is further delayed by consuming it at a later time. Each of these rewards, money, purchase of food, and eating, is connected with particular expectations. We become aware of these expectations through discrepancies. If we get paid less than we expect, we notice the discrepancy and perhaps complain to our employer. Or if we go into a shop expecting to buy an apple but find the price of apples has suddenly gone up, we notice the higher cost. And if we get home and find the apple tastes more sour than expected, we notice a further discrepancy. If everything goes to plan and meets our expectations then we may hardly think about these events. It is through discrepancies that the world announces itself to us.

These discrepancies are related to predicting overall consequences in the future. We will probably feel more disappointed if our salary is cut than if the price of apples increases; a fall in income is likely to be a greater hardship than the higher price of a few apples. Discrepancies are measured against the sum of future rewards—a salary change is weighed against the total goods we might be able to purchase with that amount of money, not a particular purchase. This property, of learning about the sum of future rewards, is also a natural outcome of TD-learning and reflects the relational way it works. Discrepancies at any given time relate to changes in expectations and rewards that follow, allowing the sum of future events to influence learning.

Learning involves continually responding to discrepancies and adjusting expectations. If our salary increases, we initially feel positive because our reward exceeds expectation. But we soon learn to expect this new salary and would feel disappointed if it returned to its previous level. The same applies to changing prices, or the quality of the apples we buy. These events are perhaps more sophisticated than Pavlov's dog salivating, or the monkey reaching for food, but the neural foundations are the same. We learn from discrepancies and adjust our expectations accordingly, shifting discrepancies in the process.

It is this relational aspect of learning, the continual shifting of discrepancies, that keeps our neural journey on the move. If we could somehow eliminate all discrepancies then learning would come to a halt. Suppose we could see perfectly into the future and anticipate exactly what was going to happen. There would no longer be any surprises or discrepancies to worry about. In this situation there would be nothing for us to learn. This might sound like an ideal state to be in. Given the way our brain works, however, it would be mentally numbing. Without anything to surprise us there would be nothing to grab our attention, no discrepancy to exercise our mind. A perfect knowledge of the future would act like a mental anesthetic.

In practice there is little risk of achieving such omniscience. This is because in spite of our rich and complex brains, the world around us is far richer. We can only hope to capture a tiny fraction of the surrounding complexity, so there will always be discrepancies and problems to deal with. That is what keeps sibyls, prophets, and fortune-tellers employed. Learning never completely eliminates challenges, it just introduces new ones.

Discrepancies are so important to us that we even create them for our own entertainment. Movies or books often end as soon as everyone is happy—without any remaining discrepancies there is nothing to keep us engaged. This seems a shame because having experienced a dramatic story, we might like to savor happiness at the end. But there is nothing more to hold our attention when all discrepancies in the story have been resolved, so happy endings tend to be much shorter than the story that precedes them. Tolstoy makes a similar point in the opening line of his book *Anna Karenina*: "Happy families are all alike; every unhappy family is unhappy in its own way." It is the discords that grab our interest.

The same principle applies to teaching. For a pupil to learn a new idea, there should be a discrepancy to latch onto. If a pupil feels they already understand or doesn't see that there is a problem to address, then little learning will happen. A good teacher has to be able to trigger discrepancies for a pupil by posing a problem or question that excites or engages them. Once a pupil has understood a particular problem by resolving that discrepancy, the next stage of learning is to introduce a further discrepancy, another problem that needs to be resolved.

A Recipe for Learning

We each pursue a particular neural journey from the time we are born. The brain of a newborn already carries a series of expectations built into its neural connections and interactions. It also has a set of neural values, ways of responding to stimuli that act as rewards or punishments, reflecting its early location in neural space. This location depends on the intricate way the neural tissue of the embryo grew and developed, on how the brain with its complex structure was formed. As the newborn starts to experience the world, discrepancies are encountered and lead to a shift in expectations, driving it into new regions of neural space. This in turn changes the neural context, introducing new discrepancies, as well as new values connected with the previous ones. And so the process continues, with the neural journey continually spurring itself on. The neural journey for each of us is unique as a result of our different early locations in neural space and the particular experiences we have. But the same set of repeating processes drives the journey. This is the principle of recurrence, but now operating in the domain of learning.

For evolution, recurrence arises from the relational way in which natural selection works. As adaptations spread and are shared in the population, all individuals are able to perform better, and this leads to competition for even more effective adaptations. For development, recurrence involves the ever shifting patterns of gene activity in a growing embryo. As a pattern is modified, it shifts the molecular and cellular context, which leads to another set of gene activations and transformations. For learning, recurrence arises in yet another form. As some neural discrepancies are resolved, additional ones are established that then drive further rounds of learning. In all cases, it is the principle of recurrence that keeps the journey going.

Like evolution and development, learning is built on a set of common interacting principles, although they are now in a neural guise. The core of learning involves a double feedback loop with neural firing reinforcing itself while also bringing about its own limits through competition. These loops are fed by variation in firing rates brought about by a population of experiences, leading to persistent changes in synaptic strengths and numbers. This happens through cooperative interactions among multiple neurons that collectively create a vast space of possible combinations. Each step in learning leads to a displacement in neural space, recurrently building upon the previous neural context, and thus modifying the brain and shifting the context. The same set of principles and interactions operate at the heart of learning as in evolution and development. Learning is our third manifestation of life's creative recipe. But instead of propelling organisms through genetic or developmental space, this instance of the recipe takes organisms on a journey through neural space.

EIGHT

Learning through Action

WE HAVE SEEN THAT like evolution and development, learning is a manifestation of life's creative recipe. My account of learning could stop at this point since we have covered the fundamental principles involved. However, to end here would leave many questions hanging. How are the principles we have covered connected with processes like memory, movement, attention, recognition, language, creativity, and consciousness? Each of these cognitive topics has been the subject of intense investigation, but unfortunately we still don't know many of the neural details of how learning is incorporated within them. Even so, the principles we have encountered can provide a framework or conceptual toolbox for understanding how learning may operate in such cases. Viewing cognition in this way allows us to explore some of the further ramifications of learning, and also sets the scene for looking at cultural changes in later chapters.

We cannot hope to cover all areas of cognition, so I have chosen to concentrate on a few key aspects to illustrate some of the main principles involved. In this chapter we look at how learning and action are interwoven, while in the next chapter we explore how we learn to see things in particular ways. In what follows I often use examples from vision, but the general principles also apply to other sensory modalities. By looking at cognition through these principles, we will see how our neural journeys are not driven passively, but are intimately connected with our actions and interpretations.

Given our lack of direct knowledge in certain areas, I use some hypothetical schemes to illustrate possible underlying mechanisms. Although this approach introduces an element of speculation, it has the merit that by following particular ideas through, we can gain a clearer appreciation of the underlying principles involved. Nevertheless, some readers may prefer to skip some of the more detailed explanations in these two chapters. Let's first turn to calibration, one of the most basic ways in which we learn to connect actions with their effects.

Calibration

Leonardo da Vinci was familiar with military conflict. During his life, Italy was under continual threat of invasion from France, and his home town of Florence was regularly at war with its neighbor, Pisa. So, it is not surprising that Leonardo often turned his inventive mind to machines of war. As well as devising military inventions, like a multibarreled machine gun, Leonardo was interested in the science of projectiles, or ballistics. He asked questions about the hurling power of a bombard, a type of cannon: "If with four pounds of powder a bombard hurls a ball of four pounds two miles, how far will six pounds of powder hurl it?" Answering this question was not only of theoretical interest but of practical importance. If you wanted to hit a military target using a bombard, you needed to know how much gunpowder to use.

There are several ways of tackling the ballistic problem Leonardo posed. A practical approach would be to record how far the ball was hurled by a bombard fired with various amounts of gunpowder. This would give a calibration table, a list of distances for each amount of powder used. You could then determine how much gunpowder was needed for a required distance by simply looking it up in the table. One limitation of this approach is that you could only look up distances that had already been covered by the tests. If you had a distance that was not listed in the table, you would be stuck. An improvement would be to construct a calibration curve, a plot of distance traveled against gunpowder used. You could then look up how much gunpowder you needed for any required distance by looking at where it intersected the curve.

Calibration is a basic example of how we may determine the action needed to attain a desired effect or goal—with a calibration curve, we

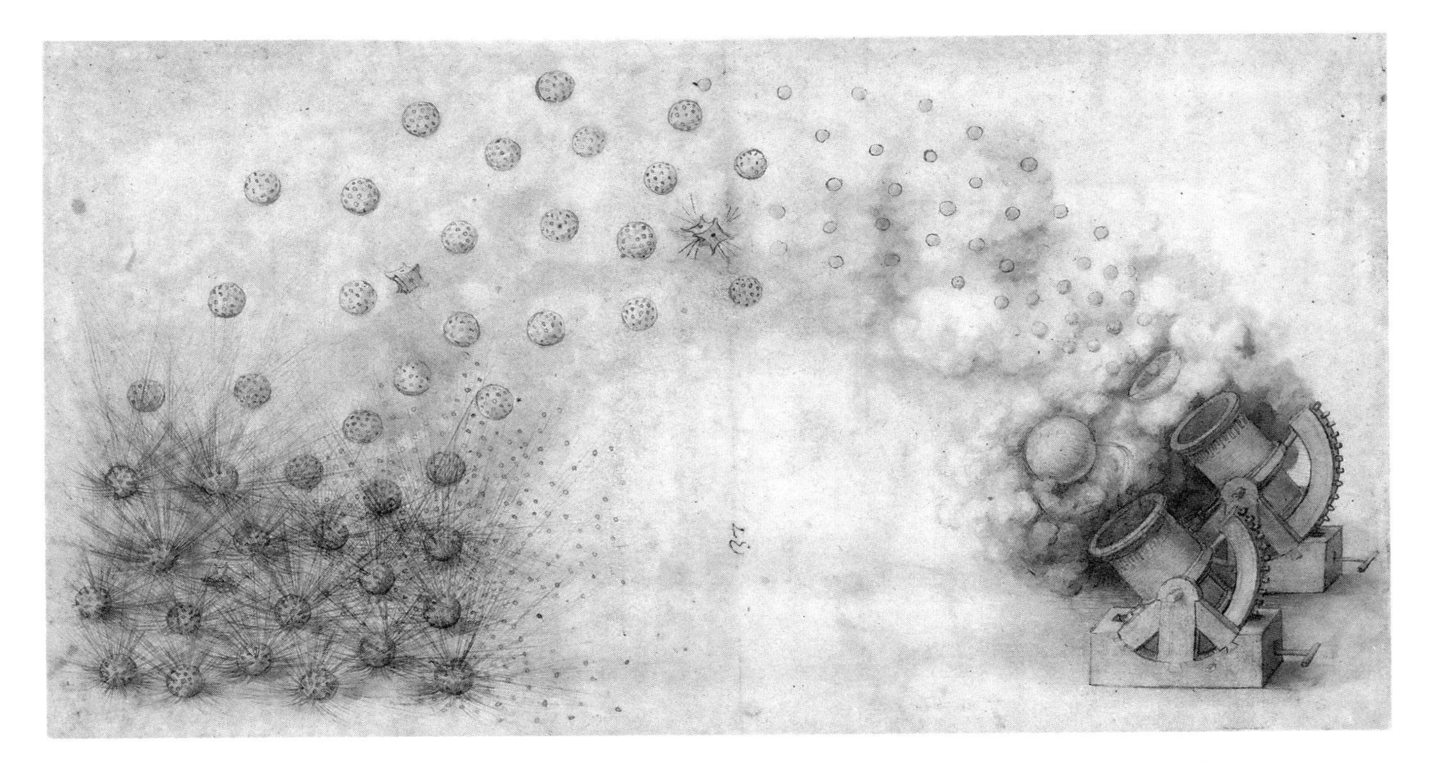

Figure 57. Bombards drawn by Leonardo da Vinci, c. 1504.

can determine how much gunpowder we need to fire the ball a desired distance. Connecting actions with goals is also fundamental to learning. In the previous chapter I described how a monkey learns to respond to the sound of a door and retrieve a piece of apple. This is a case of instrumental conditioning, where the monkey not only learns that the door sound is a predictor of food, but also that the action needed for getting that food is to reach into a box. One way the action could be learned is by the door sound reinforcing synapses involved in triggering the muscular contractions that preceded food delivery. That way, those same muscular contractions and arm movements that occurred just before the monkey got the food would be triggered every time the door sounds. But such a system would be very inflexible. If the box containing the apple reward was moved to a slightly different position, those same muscular contractions would make the monkey miss the box. In practice, if the monkey sees the box has been moved, it adjusts its actions and retrieves the reward correctly on hearing the door sound. Rather than particular muscular contractions, it seems that the monkey learns actions that relate to particular goals. If the box shifts, the goal of getting the apple is still the same, and the monkey adjusts its movements accordingly.

How might animals learn goal-related actions? A good place to begin trying to answer this question is by considering the calibration of body movements, for this is one of the most basic ways in which actions become related to their effects. Our first step is to perform the neural equivalent of calibrating a bombard.

Jumping Eyes

The most frequent action you perform in life is to move your eyes. When you look at a scene, your eyes continually jump from one place to another. These ballistic eye movements, or saccades, for someone viewing Cézanne's *Still Life with Kettle* are shown in figure 58 (plate 9). During this short time the eye leaps around many times, most often landing near objects, like the apple or vase rim. These eye movements are more frequent than heart beats—we make about three saccades every second, amounting to several billion in a lifetime. Some saccades involve relatively large jumps, like shifting from the apple to the vase; while others involve minute jiggles around a single area (called microsaccades). We are

Figure 58. Eye movements tracked for twenty seconds while viewing *Still Life with Kettle*, Paul Cézanne, c. 1869. See plate 9.

hardly aware of most saccades—they occur unconsciously through activities in our brain. Saccades are not just a feature of human vision; they are also exhibited by many other animals, from squid to goldfish.

The main reason we perform saccades involves a limitation of our eyes. There are many more photoreceptors concentrated in a central region of the retina, known as the fovea, where vision is much sharper than at the periphery. By actively moving the eye around and shifting our gaze, we bring different areas of a scene before us into high definition, providing more information about the scene. One consequence is that the image on our retina is continually changing while we look at a fixed scene like a painting. As far as the retina is concerned, the situation is the same as if someone was continually jiggling the painting while we look at it with fixed eyes. Yet we do not perceive a jittery picture as we move our eyes over it. Why is this?

Part of the answer is that we do not detect visual motion when our eyes are in the process of jumping from one position to the next. When you look at your left eye in the mirror and then shift your gaze to you right eye, you can't tell that the right eye has moved a little during the process. Your ability to detect motion seems to be suppressed while the eye is in the process of jumping. This suppression accounts for why we may be unable to sense the effect of small eye movements. But for larger saccades something else comes into play—calibration. When you initiate larger eye jumps, say from the apple to the vase in Cézanne's painting, your brain somehow knows what to expect when the eyes land. If what you see meets these expectations then you sense no motion. But if someone were to jerk the painting while your eyes are in flight, you would get an unexpected view when your eyes arrive, and would sense that there is something amiss. The situation is rather like firing a bombard that has been calibrated. If all is in order then we can correctly anticipate where the ball will fall. But if someone were to shift the landscape while the ball is in flight, the ball would land in an unexpected place and we would know something unusual had happened.

Evidence that we calibrate our eyes through learning comes from the following experiment. If you sit someone in a dark room and ask them to follow a spot of light as it jumps from one position to another, their eyes follow the target. Each time the target shifts, the eyes detect the change and fly off so that they land with the new target position in center view. These eye movements can be monitored electronically, allowing a visual trick to be performed. As the eyes fly off to follow the target, the target is moved back a little before the eyes land. Initially the eyes overshoot the target as you might expect. But after a few trials the eyes automatically adapt and learn to make slightly smaller jumps; they then start to hit the target correctly again, even though it is being shifted back. The eyes recalibrate themselves to the new situation.

Learning to calibrate makes sense because each of us is born with slightly different bodies and muscular strengths, and these may also change as we grow up. For example, the power of your eye muscles may be slightly different from mine. If our calibrations were inborn and fixed then it is difficult to see how they could work reliably when faced with such variability. In the same way, if each bombard performs a little differently because of variation in their manufacture, it would make sense to calibrate each one individually according to its performance.

Visual Shifts

What sort of neural interactions might underlie visual calibration? Let's start with a stationary scene and see how our own actions, the movements of our eyes, are related to the visual effects they generate. Imagine you are looking at a gray object on a black background with your head still. The top panel of figure 59 is a depiction of what happens at the back of one eye. To keep things simple I have represented everything in one dimension. The image of our object arriving at the retina is indicated by the gray line, while the retina is represented below this as a single file of photoreceptors, or neurons that are triggered by light. I have placed the image of the gray object in the middle of the visual field. You can see that only those photoreceptors that receive light from the object have been strongly triggered, indicated by the vertical lines inside the photoreceptors.

The bottom panel shows what happens if we shift the retina to the right with an eye movement. This movement is driven by a neuron, shown in dark gray, that triggers contractions of some eye muscles. We will assume that the greater the rate or duration of firing of this eye-movement neuron, the more the retina shifts to the right. (For the sake of simplicity I am ignoring the optical inversion that takes place when light passes into the eye.) When the retina moves to the right, there is a change in the activity of its photoreceptors. Some photoreceptors that were previously in the dark now receive light from the gray object, while others that were stimulated by light are now in the dark. Our first step is to try to relate this change in visual input to the extent of the eye shift. We want to find visual neurons that respond in proportion to the degree of eye movement.

Consider a single photoreceptor toward the left of the retina, highlighted with a white arrow in figure 59. This photoreceptor was originally silent but was triggered when the retina shifted, because the gray object came into its view. Is the activity of this photoreceptor proportional to the degree of eye shift? The trouble is that this photoreceptor is triggered to the same extent for shifts bigger than the one shown. Although the altered activity of this photoreceptor can tell us there has been some sort of

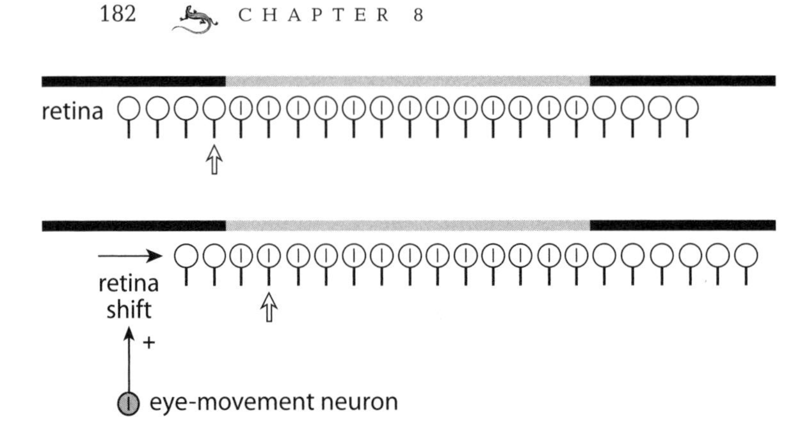

Figure 59. Object (gray horizontal line) observed by retina before (*top* panel) or after rightward shift of retina driven by an eye-movement neuron (*bottom* panel). Length of vertical lines within photoreceptors indicates level of neural activity.

change, the activity is not proportional to the extent of the eye shift. The same is true of any other photoreceptor—a single photoreceptor on its own cannot respond in proportion to the degree of eye movement.

To get around this problem we may combine information from several photoreceptors, as shown in figure 60. In the upper panel you can see three neurons, A, B, and C (ABC), each of which receives signals from several photoreceptors. All of the connections involve stimulatory synapses (positive signs). The inputs for neuron A come from the left third of the retina, the inputs for B from the middle, and the inputs for C from the right third. This means that each neuron essentially looks through a window (known as its receptive field). The window for the central neuron, B, lies over the central part of the gray object, so all of its inputs are triggered. The overall activity for neuron B is therefore high. The windows for the neurons on either side, A and C, cover both gray and black areas of the image, so they show intermediate overall activity (indicated by the smaller vertical lines within the circles).

If the retina shifts slightly to the right, as shown in the bottom panel, the activity of neuron A increases. This is because its window starts to cover more of the gray object. The activity of B remains unchanged, while the activity of C drops because more of the black background enters its window. The greater the rightward retinal shift, the greater the increase in A and the drop in C (for shifts within certain bounds). Conversely, if

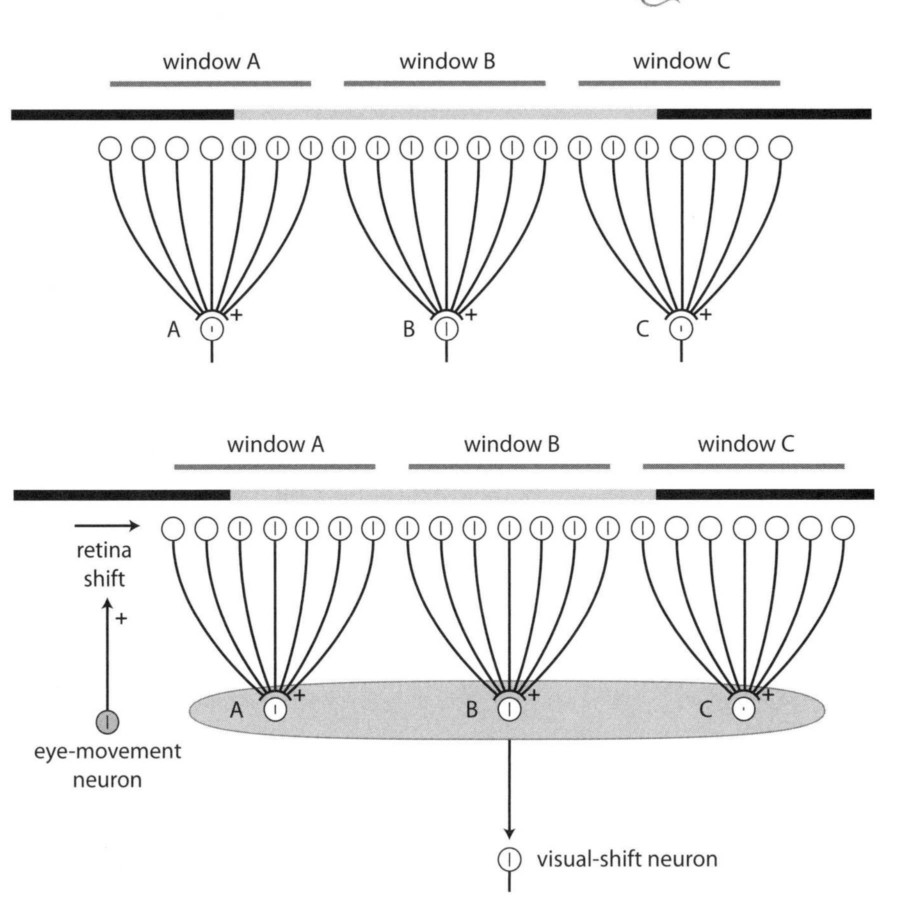

Figure 60. Combining visual information from photoreceptors. In the lower panel the retina has shifted rightward through firing of an eye-movement neuron. Length of vertical lines within photoreceptors indicates level of neural activity.

the retina were to shift to the left, the activity of A would drop while C would increase. The firing patterns of the A and C neurons can give us a measure proportional to the eye shift—the firing of A correlates positively with rightward shifts while the firing of C correlates positively with leftward shifts. Detecting visual shifts is a cooperative effort, requiring the interplay of many neurons.

Let's assume that our brain can use information of the kind provided by the ABC neurons to trigger what I call a visual-shift neuron, shown in the bottom panel of figure 60. We can arrange things so that if there has been no change in retinal input, the visual-shift neuron does not fire. The more the retina shifts to the right, the more the visual-shift neuron fires. We also assume that our visual-shift neuron behaves in this way without regard to the type of image presented. I gave an example of looking at a gray object against a black background, but we are going to suppose that the visual-shift neuron responds similarly for many other images. Shown a particular image, our visual-shift neuron fires according to how much that image subsequently moves along on the retina. Exactly how this is achieved need not concern us here; what matters is that the combined visual information from neurons like the ABC neurons is somehow used to generate a signal that correlates with the eye shift. (Strictly speaking, the signal is responding to angular shifts, because firing of the visual-shift neuron depends on the angular displacement caused by rotation of the eye.)

Learning to Calibrate

We now have two correlated changes—firing of the eye-movement neuron and firing of the visual-shift neuron. This provides a link between the worlds of action and perception. The more the eye-movement neuron fires, the more the eye shifts to the right (action), which in turn changes visual input and leads to increased firing of the visual-shift neuron (perception). I now want to show how such as system could be calibrated. Instead of constructing tables and drawing lines as for the bombard, the calibration is achieved through interactions among neurons.

Look at figure 61 which shows the same scheme as before, but with an additional discrepancy neuron. Recall from the previous chapter that discrepancy neurons are involved in the detection of discords or mismatches. To see how this could help with eye movements, let's first look at what the eye-movement neuron is doing in our scheme. You can see it does two things: it sends a signal to drive eye movement and sends the same signal (known as the efference copy) to the discrepancy neuron. The signal to the discrepancy neuron is delayed on its journey so that it arrives after the eye movement has happened. It then stimulates the discrepancy neuron via a strong stimulatory synapse (positive sign).

The signal from the visual-shift neuron also goes to the discrepancy neuron, arriving there at the same time as the delayed signal from the eye-movement neuron. However, in contrast to the eye-movement neu-

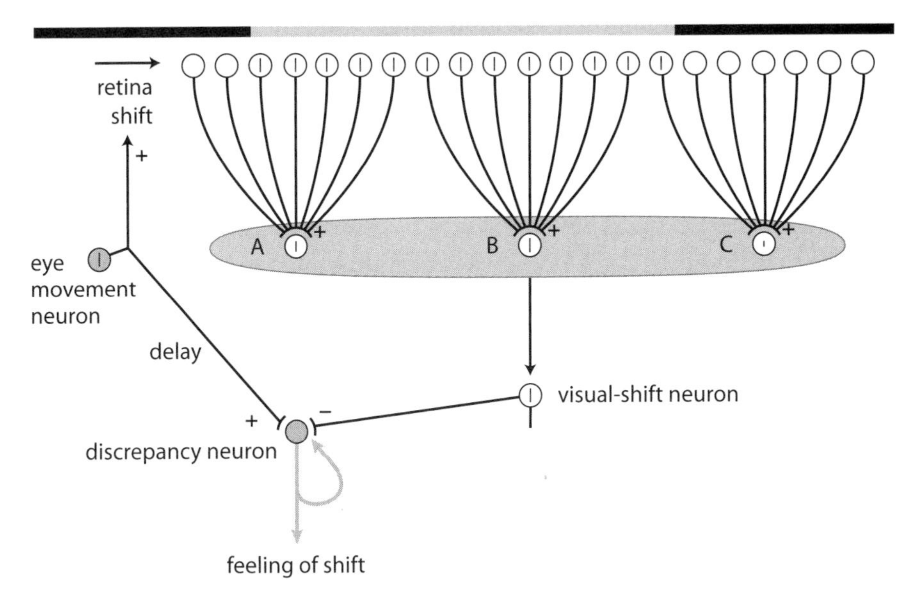

Figure 61. Learning to calibrate with a discrepancy neuron.

ron, the visual-shift neuron is connected to the discrepancy neuron by an inhibitory synapse (negative sign), so it silences the discrepancy neuron when it fires. Moreover, the strength of this synapse with the visual-shift neuron is not fixed but can change. This change depends on feedback from the discrepancy neuron (upward gray arrow). If the discrepancy neuron fires above its baseline rate then it tends to strengthen the synapse. Conversely, if the discrepancy neuron fires below its baseline rate it weakens the synapse. We have a feedback system, with the output of the discrepancy neuron influencing the strength of its input, the synapse with the visual-shift neuron.

What happens if we move our eye to the right over a fixed scene? After the eye-movement neuron fires, the visual-shift neuron also fires proportionately. Suppose the initial strength of the synapse with visual-shift neuron is weak. This means that the discrepancy neuron is poorly inhibited by the visual-shift neuron, but strongly stimulated by the eye-movement neuron. The discrepancy neuron therefore fires above its baseline level, strengthening the synapse with the visual-shift neuron. If our eye returns to its original position and then does another rightward saccade, inhibition from the visual-shift neuron is now better able to counteract stimulation from the eye-movement neuron because the synapse is stronger. If the level of inhibition is still too weak to offset the stimulation, then the strength of the inhibitory synapse is further reinforced. Reinforcement continues in this way until the inhibitory input from the visual-shift neuron comes to balance the competing stimulatory input from the eye-movement neuron. Once a balance has been established, the discrepancy neuron fires at its baseline rate and the strength of the synapse no longer changes. We have calibrated the system, matching the strength of input from the eye movement with the visual shift it generates. This took place through our familiar double feedback loop between reinforcement and competition, fueled by a balance of variation (eye movements and visual changes) and persistence (lasting changes in synaptic strength).

After our eye has been calibrated in this way, the system can tell whether a visual shift is caused by eye movement or an external event. If our eye moves to the right, the discrepancy neuron fires at its baseline rate because of the previous calibration. But suppose the same visual shift on our retina is caused by an external event that shifts the scene, as when the painting was moved. The discrepancy neuron receives no signal from the eye-movement neuron (the eyes have not moved) yet it still receives an inhibitory signal from the visual shift neuron. The overall result is a drop in firing of the discrepancy neuron below its baseline rate. Altered firing of the discrepancy neuron tells us that the event has not been caused by our own eye movement, but by some other cause.

To complete the story we need the discrepancy neuron to influence our sense of whether the scene before us is stationary or not. This is indicated by the downward gray arrow in figure 61, which links the output of the discrepancy neuron to the sense of visual shift. If the discrepancy neuron fires at its baseline rate then there is no feeling of a visual shift—the scene seems stationary. This is what happens when we perform eye movements. But if the scene shifts through external causes, then the discrepancy neuron fires differently from its baseline rate, and we sense an external shift. I have provided an example using right eye movements, but similar schemes could be devised for leftward, upward, or downward movements. In all cases we learn to calibrate our actions against the effects they cause, and then use discrepancies to infer that external events have taken place.

Such schemes account for a curious visual effect. If someone's eye movements are paralyzed with a small dose of the drug curare, the person will report that the scene in front of them tends to jerk around when they try to move their eyes. The eyes do not actually move because they are paralyzed, yet the patient senses displacement of their surroundings. In these situations the eye-movement neurons are activated but are unable to affect the paralyzed eye muscles. So, the eye-movement neurons fire without a change in visual input. The resulting imbalance would make discrepancy neurons fire differently from their baseline rate, giving us a feeling that the scene has been displaced.

I have given a very simplified scheme to highlight how our familiar principles can be related to the way eye movements are calibrated against the changes in visual input that they cause. The detailed neural mechanisms through which this calibration occurs are still poorly understood. It is known that there are regions of the brain that are organized as maps that relate visual inputs to eye movements, similar to the sensory and motor maps described in chapter 6. However, the mechanisms by which synaptic strengths or numbers are modified through experience so as to ensure that eye movements are matched to changes in visual input are less clear. Whatever the mechanisms may turn out to be, it seems likely that they will involve feedback of the kind I described, in which synaptic connections become modified through discrepancies between actions and their expected effects.

Action-Learning Loops

We often think of action as something that follows perception and learning. We perceive, learn, and then act accordingly. But the story of eye calibration shows how perception, learning, and action are much more interwoven. When we perform an action, such as moving our eyes, what we perceive may change as a result of that action—the visual scene shifts over our retina. Similar sensory changes may also be caused by external events. One of the first things the brain has to decipher is how to distinguish sensory changes brought about by our own actions from those brought about by external processes. Without doing this, we would confound movements in our surroundings with the results of our own movements. The problem can be solved by learning to calibrate our actions against the effects they produce. We can first learn what to expect from our own actions. If we detect discrepancies from this expectation, we may then infer that a change has taken place through external causes.

As illustrated by our neural scheme for eye movements, the principles involved in this process of calibration are no different from what we have encountered before. We have the interconnected feedback loops of reinforcement and competition, involved in the operation of discrepancy neurons. These loops are fueled by a balance of population variation and persistence—various actions and experiences lead to lasting modifications in synaptic strengths or numbers. And for this to happen, inputs must be brought together in close proximity so that they can act in cooperation. The overall result is that we learn how to match effects to the actions that generated them, and can then discriminate between events that are caused by our own activity from those caused externally.

The process of calibration has a further benefit—it may help guide our actions. As described earlier, if you are sitting in a dark room and a spot of light jumps to a new position, your eyes are able to follow it. According to our neural scheme, initial movement of the target creates a visual shift that in turn changes the firing of the discrepancy neuron (the eyes have not moved yet). The further the target is shifted, the more our discrepancy neuron changes its firing rate. The discrepancy neuron gives us information about how far the target has moved. We can use this information to help guide our eye movements. To achieve this, the output from discrepancy neurons can be fed back to move our eyes. The principles are the same as before, except that we are now using discrepancies not just to detect external events, but we are also feeding the information back to affect our actions. If the target is surreptitiously moved back during saccades, the system automatically recalibrates and modifies the extent of eye movement accordingly. Calibration not only enables us to distinguish internal from external events; it helps guide our actions in response to external change.

Smooth Movements

We have seen how calibration can be applied to ballistic movements like eye jumps. But many of our actions are much smoother than this. We do not walk like jerky robots, but have a much more fluid set of motions available to us. How do we learn to control smooth actions?

Modern weapons are much more elaborate than those of Leonardo's time. Instead of firing a missile and hoping that it lands in the right spot, we may fire guided missiles that lock onto a target. These missiles have sensory systems, such as built-in radar, that detect errors between their trajectories and the position of the target, and adjust their course accordingly. Control systems of this kind have the advantage of being able to adapt if the missiles are knocked off course by external factors, such as the wind; they also have the flexibility to follow moving targets. Equivalent systems are involved in our bodily actions. As before, I want to use our visual system to illustrate some of the principles involved.

If you look at an object and slowly turn your head, your eyes will stay locked onto the object. Your eyes do this for several reasons. One involves neurons in your eye and brain that are able to detect the velocity at which images move over the retina. Like the visual-shift neurons I described earlier, these visual-velocity neurons operate by comparing signals from different photoreceptors of the retina received at various times. But they operate on a shorter timescale—while the eye is stationary or moving slowly, rather than between eye jumps. As your head moves, the visual-velocity neurons detect the speed and direction with which the image is moving over the retina. This information is then used to move the eyes. If the velocity of the eyes matches the initially detected velocity of the image, then the image is once again stationary on the retina, and the object is held in view. But if there is any mismatch, then the image starts to slip over the retina to some extent, triggering the visual-velocity neurons again and leading to a further adjustment in eye velocity. In this way our eyes use a continuous feedback system to keep track of the object, much as a guided missile stays locked onto its target. Unlike saccades, this type of eye movement is smooth, so your gaze can be held in place as you gradually turn your head.

There is another system that helps keep your eyes on target. Even with your eyes closed, they tend to stay oriented toward the same external position while you turn your head. There is a natural tendency for them to rotate in the opposite direction of the head and counter the head movement. This response depends on neural receptors in the inner ear that detect head movement and send this information to influence eye movement. As with the visual-velocity neurons, this system counteracts head movement with a smooth eye motion and stops the image from slipping over the retina too much.

We have two systems that keep us on target, one that senses visual-velocity and the other that senses head movement. How are these two systems integrated? As we will see, the answer is that one is calibrated against the other.

In the 1890s, George M. Stratton carried out what has been described by Richard Gregory as perhaps the most famous experiment in all of experimental psychology. He wore up-down and sideways reversing goggles for several days to determine whether his vision would adapt to the new situation. These goggles reverse the effect of head movements, so when you turn your head one way, the visual field goes in the opposite direction to what would normally happen. When he first put on these goggles Stratton felt very confused—turning his head gave him the opposite visual sensation to what he expected. But after wearing the goggles continuously for about a week, he found that things appeared relatively normal again. He had adapted to the new arrangement. Later studies showed that one effect of such adaptation is that the response to head movement is reversed—the eyes turn in the direction opposite to how they normally would when the head turns. This happens even in the dark. Although moving in the opposite direction, the overall effect of the eye movements is the same—to reduce slippage of the image on the retina, though now with the reversing goggles on. How does this readjustment happen?

As we turn our head from left to right, neurons in regions of the brain, such as the cerebellum, learn to calibrate inputs from the inner ear and visual system. These neurons may operate like the discrepancy neurons we encountered for saccades, balancing one input against another. After calibrating ourselves, if we turn our head with our eyes closed, the discrepancy neurons in the cerebellum fire, because there is no visual signal to counterbalance the input from the inner ear. This firing of the discrepancy neurons would lead to the eyes turning by the appropriate amount. After wearing reversing goggles for a while, the discrepancy neurons recalibrate to the new situation, capturing the reversed relationship between signals from the inner ear and visual-velocity neurons. The result is that when the head turns, the eyes are driven in the opposite direction than before, helping the gaze to be maintained in the new situation.

As well as helping our eyes stay on target, such calibration helps to distinguish the effects of our movements from those of the surroundings. When George Stratton first put reversing goggles on, he had a sense of external motion when he moved his head: "It did not feel as if I were visually ranging over a set of motionless objects, but the whole field of things swept and swung before my eyes." This is because his brain had earlier been calibrated on normal vision, and was now detecting discrepancies between his actions and their expected visual effects. According to our scheme, particular discrepancy neurons were firing and triggering his feeling of external motion. After wearing the goggles for a few days, his discrepancy neurons recalibrated, so the feeling of motion that followed his actions had disappeared.

The advantage of calibrating ourselves in this way is that we are able to discriminate external events from movements caused by our own actions. If you look outside the window of a moving train you feel the world is moving relative to you. Discrepancy neurons may be firing because you are getting signals from your visual-velocity neurons, while there is no balancing signal coming from neurons that sense or drive head movements. The resulting discrepancy leads you to sense external motion. Our feeling of motion is relational, and is based on how sensory inputs deviate from what our own actions lead us to expect.

I have emphasized how we calibrate our actions against visual inputs, but the same applies to inputs from other senses. If you stand next to a man in a dark room and stretch out to touch him, you attribute the sense of pressure against your hand to your own action. The man is fixed and you have touched him. Alternatively, if you hold out a stationary arm and suddenly feel the man's body pressing against your hand, you assume that he has moved, and not you. The sensation in your hand may be exactly the same in both cases, yet only in the second do we sense external movement. This is because we have learned to calibrate our bodily actions against the effects they cause. Through many previous acts, beginning with our random flailings as babies, particular discrepancy neurons may have learned to settle down to baseline activity levels according to the effects of our movements. Only when sensations deviate from this expectation do we sense external motion.

All of these types of calibration come into play while walking. As you walk along through a still landscape, you sense that all motion has to do with your own actions. Having previously calibrated the signals from your feet, legs, eyes, and head with each other and with neurons that drive their activities, you are aware that you are walking through a fixed world. By contrast, if you are standing on a moving walkway, you sense that you are being moved along. Even though the visual information you receive is similar to what you would acquire by walking, neurons that normally fire when you move your legs are silent. This deviation from expectation may trigger particular discrepancy neurons, giving a sense of externally driven motion. If you start to walk on the walkway, then you begin to feel that the visual shifts you perceive are accounted for by your own actions. But there is still some discrepancy, because the visual shifts are greater than you would expect from your previous experiences of walking, and you have the feeling that you are able to walk much faster than you normally do.

Such calibrations are important because they enable us to anchor ourselves in the world and then act more effectively within it. If you see an apple hanging from a tree, walking up to the apple and grasping it is a relatively straightforward action. Imagine, though, that you had not previously calibrated your actions with their effects. As you look at the apple, the scene before you would seem to start jumping around with your every eye movement. Turning your head toward the apple would make the world spin before you. And as you try to walk toward the tree you would sense the ground moving, and rising to meet your feet. Everything would seem to swim around you and the more you act to correct this, the worse it would get. Instead of walking smoothly toward the apple, you would stagger around like a confused drunk.

The ability to anchor ourselves in the world depends on a whole series of cross-calibrations. Through such combinatorial richness, our neurons are able to integrate multiple actions and their sensory effects. This allows us to continually learn from discrepancies by comparing one component in relation to another. By recurrently applying the same relational system, we then learn to calibrate multiple systems against each other and act more effectively. The principles of combinatorial richness and recurrence are not only important for learning from our surroundings, they are central to our actions.

An Active Journey

In the previous chapter we saw how learning could be thought of as a journey through neural space. We now see that this journey does not simply involve us responding to events in our surroundings. We continually act and by doing so, change what we experience. When we move our eyes or turn our head, we experience new sights. We can learn from the way our actions influence our experiences, and this information is then fed back to influence further actions and experiences. Our journey through neural space is not passively driven by our surroundings and sensations; it involves continual feedback with our actions.

We are born with some instinctive actions and responses already in place—you only need to see a newborn child wriggling and throwing its arms and legs around to appreciate this. These instinctive reactions provide the launching pad for a much longer neural journey during which our actions and experiences continually feed off each other. Through this process, we learn relationships between actions and their effects, and this drives us into new regions of neural space. Calibration provides an illustration of how such learning takes place. But examples like this are only the tip of the iceberg.

Let's return to the experiments of Romo and Schultz in the previous chapter, where the monkey learned to respond to the sound of a door and then retrieve a piece of apple from a box. We can break this process down into two types of learning. The first is learning the action needed to retrieve the reward—reaching into the box. The second is learning that the door sound is a predictor that this action will be rewarding. (Without the door sound, the box is closed and inaccessible to the monkey.) I now want to look at each of these learning steps in turn.

Before the monkey was trained to respond to the door sound, Romo and Schultz left the door to the box open for a while, allowing the monkey to learn that reaching into the box sometimes yielded a rewarding piece of fruit. This learning arises through self-initiated exploratory behavior of the monkey. The monkey need not wait for an external signal to explore; it begins its own exploratory movements within a given context (in this case, the monkey's cage). You might call this behavior instinctive curiosity. The monkey then finds that initiation of some actions, such as reaching into the box, are sometimes followed by rewards. In other words, it learns a goal-related action, an expectation that a particular reward may be obtained after initiating a given action. This is similar to what we encountered for visual calibration, where a particular visual shift comes to be expected after initiating an eye movement. The main difference is that for the monkey's exploratory learning, the relationship between action and effect is more complicated—only a few of the many possible actions result in rewards, and there are also variable time delays between initiation of actions and their effects. Nevertheless, the basic principles of learning through the interplay of actions, expectations, and discrepancies are the same.

By relating actions to their effects, the process of exploratory learning is not only similar to calibration; it is built upon it. The monkey's exploratory movements are only possible because the monkey has already become anchored in the world and knows how to act within it. This earlier calibration also means that if the box is moved slightly the monkey knows how to adjust its movements so as to keep finding the reward. It is through the interplay between calibration and exploration that goal-related actions can be learned.

Training with the door sound introduces a further environmental factor to this process. Access to the box is now blocked unless there is a door sound. The monkey then learns that a particular sensory experience, the sound of the door, signifies that the action of reaching into the box will be rewarding. Such behavior can arise by extending the type of learning scheme described in the previous chapter, through incorporation of goal-related actions. Rather than simply strengthening links with particular monkey movements, the sensory input becomes connected with the goal-related action of retrieving the reward. This means that if the box is moved, the monkey will still be able to perform the appropriate action.

This example involves a single goal-related action, retrieving the apple reward; but multiple goals are also possible, and these introduce further levels of cooperation and competition. Suppose the monkey is presented with two boxes containing apple rewards, one to its left and one to its right, and that they are both opened at the same time with the door sound. Through multiple trials the monkey learns that rewards are to be found in both boxes. If the monkey can only move one arm, then it is faced with a choice when it hears the door sound of reaching either to the left or the right. If the rewards in each box are similar, then perhaps it would move to the left or right with equal probability. It is important, though, that the monkey arrives at a decision; otherwise it would be stuck, unable to decide which way to go and thus get no reward at all. Potential actions therefore have to fight it out with each other in the brain so that one wins. This competition may operate along similar lines to those we encountered in development, where cells fought over the expression of a particular gene (Delta, as discussed in chapter 3 on pages 73–75). In that case, cells inhibit each other's gene activity, with the result that if the activity of one cell slightly exceeds the other, then its advantage is reinforced and it eventually fully represses the other. The difference for the monkey's decision is that the struggle takes place on a much shorter timescale and involves neurons inhibiting each other's firing. Of course, monkeys and humans are not aware that such neural competition is taking place when they make a decision. We feel we make decisions of our own accord. Yet behind the scenes, neurons may be fighting it out to ensure that eventually one action prevails.

We may also have the situation where the rewards in the two boxes are not the same. Suppose the boxes are sometimes empty, perhaps with the left box being empty more often than the right. The monkey is then faced with a degree of uncertainty about whether it will get a reward or not. Over time, it learns to favor the right box because it is more rewarding on average; but the monkey also checks the left box every so often, just in case there has been a change in the probability of finding some apple. The monkey is learning not only about rewards in relation to individual actions, but also the best policy to adopt for getting a reward given a range of choices and situations. Such learning can again be achieved through the interplay between actions, expectations, and discrepancies, although the neural schemes involved are more elaborate.

Another factor that influences such choices is the incentive, or relative value, connected to different rewards. These incentives may themselves change according to what the monkey has experienced. Suppose, for example, that the monkey gets fed up with eating lots of apple and yearns for a new food, such as banana. If we now place banana in the box on the right, then over several trials the monkey learns to go to the right more often than the left. Conversely, if the monkey had already eaten lots of banana it may prefer to go to the left to retrieve apple. The notion of a reward is not fixed but also depends on the monkey's history of experiences.

Learning enables the monkey to make informed decisions about its best options given its earlier experiences and current situation. These decisions are always relational, weighing one set of actions against another. Before banana appeared on the scene, the monkey may have been happy always to retrieve apple. Introducing another option shifts expectations and behavior. The basic principles are no different from what we encountered before. At the heart of the process are discrepancy neurons, operating according to our familiar twin feedback loops. The loops are fueled by a balance of variation (a range of experiences and actions) and persistence (lasting modification in the strength and number of synapses). The main difference is that we have introduced further levels of interaction between actions and their effects; further degrees of competition, cooperation, combinatorial richness, and recurrence.

Without this rich interplay between action and experience, our neural journeys would be very different. Suppose you were born without the ability to move, but with all of your senses intact. You would be unable to shift your arms or eyes, or make a sound, even though you could see and hear what is around you. (I assume bodily functions like breathing and the heart pumping still work to keep you alive.) In this situation, it is doubtful you would be able to make much sense of what you hear and see. If someone showed you an apple you would have little idea of what was happening. You would be unable to follow it with your eyes, and having never grasped an apple or bitten into it, you would have no idea of how it felt or tasted. Even the notion of an apple as an external object would be foreign to you, for there would have been no opportunity to actively explore the relationship between yourself and any objects around you. Not being able to move would be a far greater disability that losing one of your senses. A blind or deaf person can still form a coherent notion of the world because they are able to move around and explore it through their own actions. The importance of being able to act in order to make sense of the world was emphasized by the physiologist Hermann von Helmholtz, who wrote in 1866:

> It is only by voluntarily bringing our organs of sense in various relations to objects that we learn to be sure as to our judgments

of the causes of our sensations. This kind of experimentation begins in earliest youth and continues all through life without interruption. If the objects had simply been passed in review before our eyes by some foreign force without our being able to do anything about them, probably we should never have found our way about amid such an optical phantasmagoria.

Learning with Others

So far, we have looked at how individuals may learn through their own efforts. But particularly for social animals like humans, there is a further option—learning from others. We may learn to copy someone's actions or learn from what they do or say. It may seem that this form of learning is quite different from what I have described so far, because information seems to come purely from external sources. However, even here, learning may be grounded in our own actions.

In the 1990s, Vittorio Gallese, Giacomo Rizzolatti, and colleagues at the University of Parma were recording electrical signals from neurons in the brains of macaque monkeys. They identified a particular set of neurons that were activated when a monkey performed certain goal-related actions, like picking some fruit off a table. Remarkably, they found that these same neurons were also activated when the monkey watched the experimenter or another monkey perform the same action. The goal is important here. If a monkey watches someone just grasp, when there is no fruit on the table, then the neurons do not fire. These neurons are called mirror neurons, because they respond to external actions in a way that mirrors internally driven actions.

Mirror neurons indicate that our response to others may be closely connected with our own actions. As we learn goal-related actions, we may strengthen neural connections between particular actions and their effects. The monkey witnesses its own arm movements as it stretches out to reach some fruit, and connects this with getting a reward. When the monkey sees the same type of action being carried out by another individual, some of the same neurons may then be triggered. The overall result is that the actions of another individual become linked with the monkey's own actions. This link may then allow the monkey to interpret what another individual is up to, and so learn from them. Learning to understand or imitate others is not a one-way street. It is grounded in the way we learn from our own actions and then use this knowledge to help forge links with others. Just as an immobilized person would not understand an apple placed before them, they would have no idea what someone else grasping that apple means; because having never performed such a movement themselves, the action would seem completely mysterious.

This interplay between our movements and those of others, allows for yet another level of cooperation. If you are talking with someone, you are performing all sorts of actions. As well as speaking part of the time, you are moving your eyes, changing the expression on your face, and adjusting your posture. These actions are sensed by your partner, and elicit particular reactions. Their eyes may follow yours and their facial expression may change according to what you say and do. You in turn sense these reactions. Your actions are being transmitted back to you via your partner. Of course the same is also happening to your partner; their actions are reflected back to them through you. We have two interlocking feedback loops in which the movements of you and your partner become intertwined. This mutual cooperation enables effective communication between your brain and theirs. We can appreciate this most clearly when one of the feedback loops becomes broken. If your partner's eyes glaze over, or if they reply in a way that shows that they have not listened to what you said, then you sense that communication is blocked. Your actions are no longer receiving their expected feedback from your partner, and the intimacy between you is lost. As neuroscientist Chris Frith says: "When two people interact face-to-face, their exchange of meaning is a cooperative venture."

The neural journeys we take upon entering the world are not independent of those taken by others. Our journeys continually influence and respond to the journeys taken around us. The basic principles by which this happens may be no different from what we have already encountered. At the heart of the process is detecting and learning from discrepancies through our familiar double feedback loop. The only difference is that there are now further levels of cooperation, combinatorial richness, and recurrence involving the interplay between our actions and experiences, and those of others.

We have seen how life's creative recipe provides a conceptual framework for understanding our ability to learn from our own actions and those of others. Even so, grasping some fruit or having a conversation may seem a far cry from coming up with a new idea or creating a beautiful painting. What lies behind such intelligence and creativity, and how does it relate to the processes we have encountered? To answer this question we will look at how we learn to see things in particular ways.

NINE
Seeing As

TEN YEARS AFTER Cézanne painted his portrait of the art dealer Ambroise Vollard (plate 1), Pablo Picasso produced his own version of the same subject (fig. 62, plate 2). Painted in the Cubist style, Picasso's portrait is very different from Cézanne's. While Cézanne portrays his subject as a solid form with blocks of color, Picasso uses angular elements and a more abstract approach. By exploring different ways of seeing the subject, each artist captures distinct aspects, highlighting particular relationships in his own way.

Appreciating relationships from fresh perspectives is not only a feature of art; it lies at the heart of all forms of human creativity. Wordsworth saw his lonely wanderings in those of a cloud. Newton related the fall of an apple to planets circling the sun. Beethoven explored new tonal relationships when composing his symphonies. On a more modest scale, we perform similar creative acts every day, by activities like arranging flowers in a vase or constructing a new sentence when we speak. In all these cases, relationships are being explored in fresh ways, whether they are between objects, movements, sounds, or words.

In the previous two chapters we examined how we learn to predict and act. I now want to look at how we learn to interpret or view things in the world around us in particular ways. We will see that these three processes—prediction, action, and interpretation—are intimately connected. It is through their interplay that we will arrive at the roots of human creativity. As you might have anticipated by now, these roots do not involve any fundamentally new principles—we will find our familiar

Figure 62. *Portrait of Ambroise Vollard*, Pablo Picasso, 1909. See plate 2.

creative recipe at work. But the recipe involves further ramifications in the way neurons interact and play off each other.

A good place to begin is with ambiguity. Look at the picture in the middle of figure 63. You may either see a haggard old woman in profile or a young woman turning away. These two interpretations are made clearer in the pictures on either side, where I have introduced some modifications to bring out the old (*left*) or young (*right*) woman. When looking at the middle picture you may see one of two women, yet the sensory input is exactly the same. We encountered a similar situation the paintings of Arcimboldo (plates 7 and 8) which can either be seen as showing piles of books and vegetables, or as faces. Such examples show that perceiving is not just about what our senses detect; it also depends on the interpretations we bring. Our perceptions have a strong component that comes from us.

Our perceptions are also constrained in certain ways. When looking at ambiguous pictures such as the one in the middle of figure 63, we only make one interpretation at a time—either we see an old woman in profile or a young woman turning away, but not both at once. Our brain seems constrained to resolve the picture one way or another. Also, what we see may be influenced by what we already have in mind. If you first look at the picture on the left of figure 63 and then switch to looking at the middle picture, you are more likely to see the old woman. Conversely, if you had come to the middle image having started with the one on the right, you are more likely to see the young woman. What we see may depend on what we previously saw or were expecting. These expectations are rooted in even earlier experiences. Had we never seen old or young women, we would view the middle panel of figure 63 in a different light. Perhaps we would see it as an abstract black and white pattern, which is yet another valid interpretation. Our interpretations are not given; they are constrained by our history of experiences.

What is the neural basis of interpretations and why are they constrained in this manner? Certain regions of the brain are known to be involved in responding to particular objects. For example, there is an area known as the fusiform face area that is involved in recognition of complex objects like faces. People with lesions in this area have difficulty in interpreting or recognizing faces, a condition known as *prosopagnosia*. However, we don't yet know how such regions of the brain, or the neurons

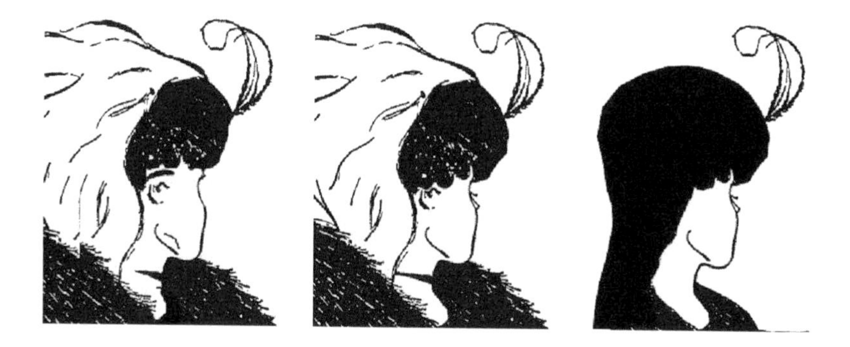

Figure 63. Middle drawing: *My Wife and My Mother-in-Law.* Edwin Boring, 1930. The drawing has been modified to emphasize the old (*left*) or young (*right*) woman.

within them, acquire their interpretive abilities, so a detailed answer to how we learn to see things in particular ways cannot currently be given. Nevertheless, we can use the principles and interactions we have covered as a conceptual toolbox to explore some of the issues involved.

When trying to traverse a largely unknown terrain, one strategy is to strike out in a particular direction to gain an idea of the lay of the land. The direction may not be the best one to take, but following it should reveal some of the problems to be faced. This is the approach I want to take in this chapter. I want to present a particular way of considering the neural basis of interpretations. My aim is not to provide a definitive account—doubtless many of the details that follow will turn out to be incorrect. Rather, I wish to pursue this path so as to highlight some of the key principles and problems involved. The going may get a little heavy at times but remember that it is not the particulars of the journey that matter as much as what we learn along the way.

I use the interpretation of visual images as my main example, although the principles uncovered are equally applicable to other sensory modalities, like sound or touch. As always, it helps to first simplify the problem. Rather than dealing with interpretation of complex images like faces, I want to begin with some simpler, one-dimensional cases.

Suppose you live in a world containing linear people. Each person in this world comprises a light gray line with an inner dark patch. Figure 64 shows two of the people you encounter. One of them is Mary, recognizable because of her central dark patch. The other is John whose dark patch

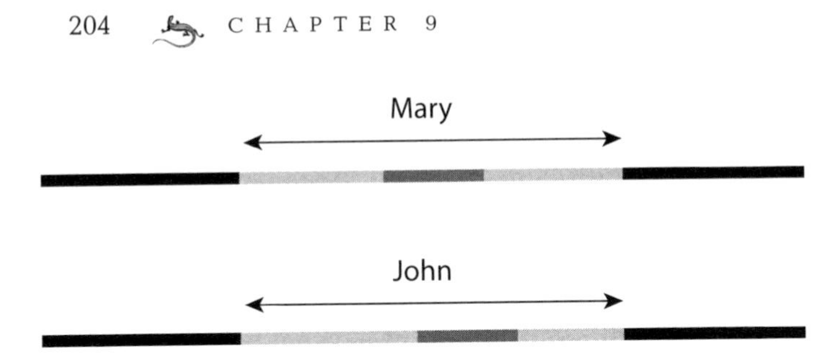

Figure 64. One-dimensional John and Mary.

is slightly off center. When looking at figure 64, we see that John is like Mary but with the dark patch shifted slightly to the right. We are able to see this relationship because we are so accustomed to looking at, and interpreting, objects around us. The problem is to devise a neural system that achieves the same result—a system that sees John and Mary as the same sort of entity, while also detecting that they are different. To help with this problem I am going to introduce what I call *neural eyes*. Although neural eyes may seem rather theoretical to begin with, we will eventually see how they may correspond to various levels of processing in the brain.

The Neural Eye

When we move our eyes, a change in eye position leads to a corresponding change in neural inputs to our brain. But there is another way of achieving the same result that involves no eye movement. The basic elements needed are illustrated in figure 65. In the left part of figure 65 you can see a neuron (the receiver neuron) receiving stimulatory inputs from a neuron to its left (L), and one to its right (R). The same arrangement is repeated in the right part of the figure but with a further neuron added, which I have called a selection neuron. Firing of the selection neuron has the effect of blocking the ability of the L neuron to send its signal to the receiver neuron, as if the connecting wire has been cut. This can happen through a process called presynaptic inhibition, in which the firing of one neuron can interfere with the ability of another to release neurotransmitters from its terminal. The result is that the receiver neuron only has a live connection to the R neuron, while the connection to the L neuron is

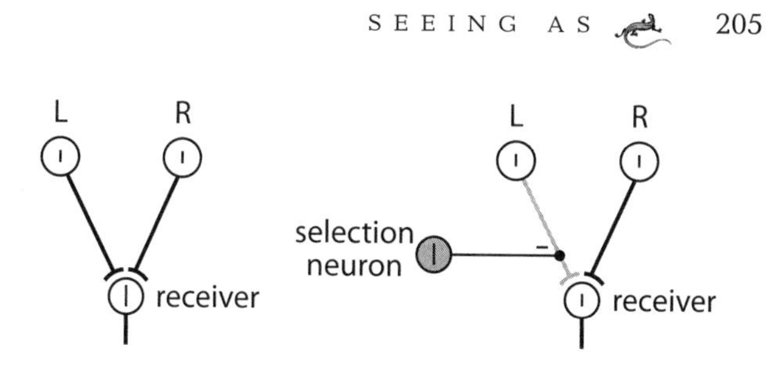

Figure 65. Selecting inputs to a neuron.

dead (shown in gray). The selection neuron has effectively changed what the receiver listens to. This change is reversible: if the selection neuron stops firing, then the connection to the L neuron revives and the receiver listens to both the L and R neurons again.

In this example, firing of the selection neuron simply blocks transmission along one line of communication. Slightly more complicated schemes can be devised in which firing of a selection neuron may shift the live connection from one neuron to another. For example, suppose initially the left connection is live and the right one inhibited, and that firing of a selection neuron kills the left connection while releasing the one on the right. In this case, firing of the selection neuron effectively moves the receiver's attention from one neuron to the other. Again this change is reversible: if the selection neuron stops firing, the live connection shifts back to the left input.

If we apply this idea to many neurons, we can arrive at the situation illustrated in figure 66. The upper panel shows the image of Mary over some retinal photoreceptors. These neurons feed their signal to another set of neurons, operating at what I call level 1. I have shown two connections between each level 1 neuron and the photoreceptors in the retina. One of these is live (black) while the other is dead (gray). The live connections point directly up to the retina. Because of the ordered arrangement of these connections, level 1 essentially mirrors the pattern of activity in the retina, forming what is known as a retinotopic map. The gray connections are similarly ordered, but are offset, and form diagonal links to the retinal photoreceptors. Although I have shown direct connections between level 1 and the retina, the connections can be much more indirect and combined

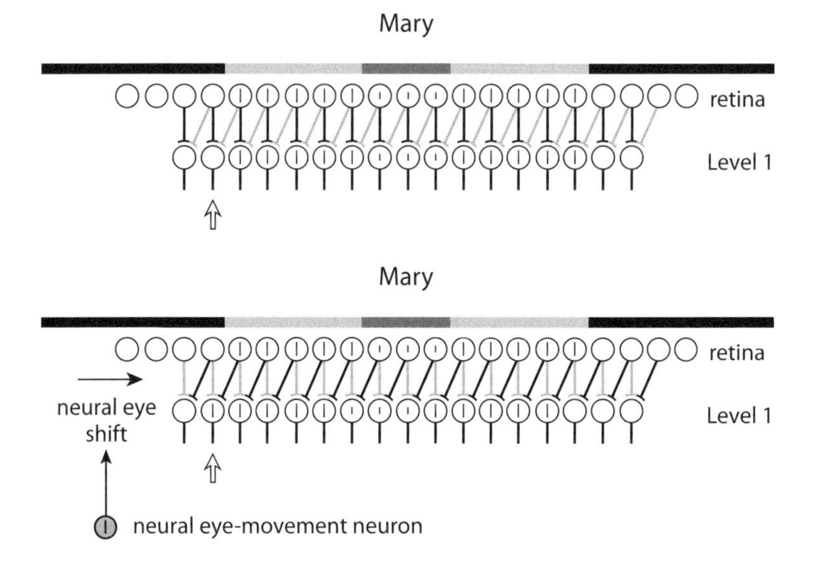

Figure 66. Observing Mary with the vertical connections to level 1 being live (*top* panel), or after a neural eye movement which involves the diagonal connections becoming live (*bottom* panel). Length of vertical lines within neurons indicates level of neural activity.

at various intermediate levels (we will see later that level 1 may correspond to a higher level area in the visual cortex of the brain). All that matters in what follows is that the various connections to level 1 are arranged in retinotopic manner, reflecting the organization of the eye.

I now want to introduce the notion of a neural eye movement, equivalent to a physical movement of an eye. Look at the bottom panel of figure 66, where I have shown what I call a neural eye-movement neuron. This neuron influences which of the connections between the retina and level 1 are live. I have arranged it so that increased firing of the neural eye-movement neuron shifts all the live connections to the right. There is no physical movement of the connections, only a change in which connections are transmitting neural signals from the photoreceptors. The previously gray connections turn black while the previously black connections turn gray. Following such a shift, each level 1 neuron receives its signal from a region further along the retina. For example, the neuron highlighted with the white arrow, which was originally getting no signal, now receives input from Mary. The overall result is equivalent to what would happen if we had retained the original connections and moved the retina to the right. The neural eye-movement neuron is doing a similar job to the eye-movement neuron in our schemes from the previous chapter (fig. 59 on page 182). Only now we are shifting which connections are live internally, rather than physically moving the retina.

In the example, inputs were shifted by one neuron to the right, but you could also imagine other shifts. For this we would need more dead connections between the each level 1 neuron and the retina, extending further along the retina to the left and right. At any one time, only one of these many potential connections with each level 1 neuron would be live. Suppose that the more the neural eye-movement neuron fires, the more the live connections shift toward the right. Similarly, firing of another neural eye-movement neuron could shift live connections toward the left. In these ways we can move our neural eye just as we move our normal eye. But there is no physical movement of the eye here, only switches in which connections are live. For convenience I shall often refer to these shifts in what level 1 receives as movements of a neural eye, as they have a similar effect to an eye movement. But remember that a neural eye movement involves no physical movement of the eye, only a change in which connections are live. Movement of a neural eye is still a physical process, but involves changes in the signaling ability of neurons in the brain (e.g. through presynaptic inhibition) rather than contraction of eye muscles.

Suppose we now learn to follow Mary with our neural eye as she moves around relative to us, left or right against her black background line (this motion may also be generated by us moving relative to her). This would involve a form of calibration similar to what we encountered in the previous chapter; we would detect discrepancies when she moves relative to us and then adjust our neural eyes to reduce them. Now, however, we are calibrating our neural eye movements rather than physical movements. We calibrate according to a particular view of Mary—when she is center stage. Our neural eye always tries to retain this original image as she moves around.

This process of calibration requires our neural system to be able to detect shifts in Mary's position. Recall that being able to detect visual shifts or motion requires combining information from several photoreceptors. A possible scheme for how our neural eye might do this is shown in figure 67 where I have introduced a further level (level 2) with three neurons, A, B, and C (ABC) that receive their inputs from level 1. The A

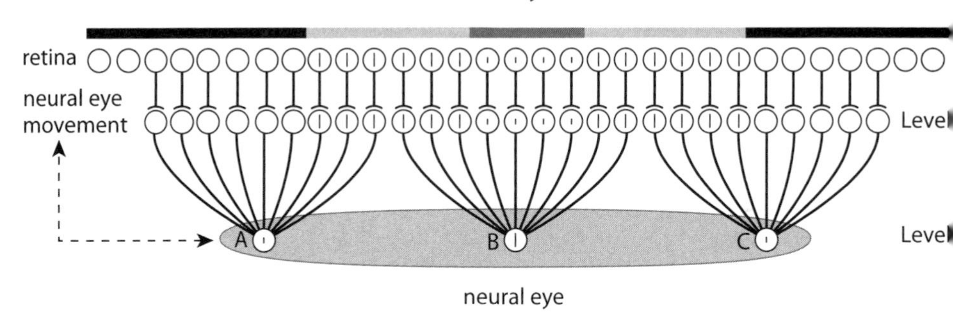

Figure 67. Connecting neural eye movements with visual signals received by the ABC neurons.

neuron looks at the left third of the visual field, the B neuron at the central third, and the C neuron at the right third. To keep things simple, I have shown only the live connections between the retina and level 1, not the many dead (gray) connections that are also present. As Mary moves around slightly from left to right over the background, we may now learn to track Mary's motion with neural eye movements, always trying to maintain the original view we had of her. Changes detected by the ABC neurons would be fed back to the neural eye-movement neuron, keeping the neural eye on track. The net result is that if Mary moves relative to us, the neural eye follows, keeping her in its sight. This has the effect of stabilizing Mary's image, at least for regions of the brain that operate in this manner.

Multiple Eyes

So far, moving a neural eye has not achieved anything we could not already do by physically moving an eye. But neural eyes have one big advantage over normal eyes—we can have more than two of them. As pointed out by David Van Essen and colleagues at Caltech in 1993, multiple neural eyes could help with making visual interpretations. To get an idea of how this might work, we have to add another level to our scheme.

Figure 68 shows the same scheme as before but with an extra level inserted. Instead of one set of ABC neurons at level 2, we now have three

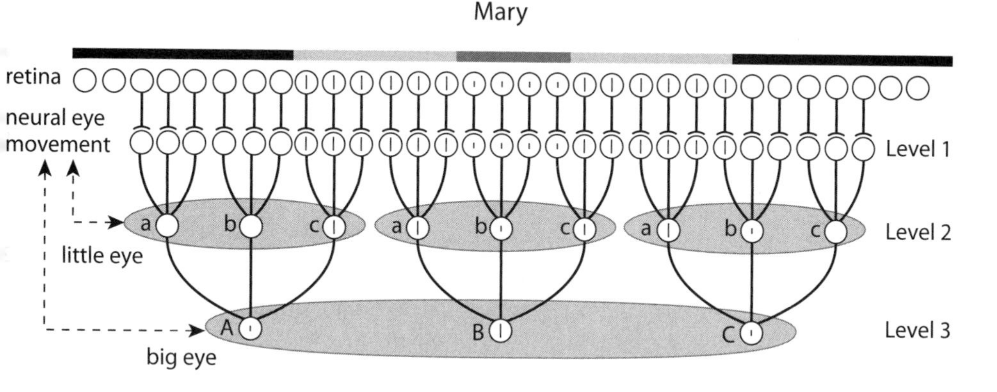

Figure 68. Neural eyes with an additional level.

sets, named a, b, and c (abc). Each of these abc sets is like the ABC set, except that they involve a smaller part of the visual field. Whereas information from the ABC neurons (now level 3) collectively spans the entire field, each abc set only spans one-third of the field.

Just as the ABC neurons described previously could be thought of as a neural eye that can be moved by controlling which of its incoming connections are live, the abc neurons at level 2 can also be considered as neural eyes that can be moved by controlling which of their inputs are live. It is as if we have four moveable neural eyes, one big neural eye (big oval ABC, level 3) taking in our entire scene, and three little neural eyes (small ovals abc, level 2) each taking in only parts of it. Each of these neural eyes has its own neural eye-movement neurons that control which of its incoming connections are live and thus effectively move the eye. (To keep the diagram simple, I have not shown these neurons individually.) Firing of the neural eye-movement neurons that control the big eye, shifts all the live connections between the photoreceptors and level 1 along. Firing of the neural eye-movement neurons that control the little neural eye on the left, shifts live connections between the photoreceptors and level 1 along for the left third of the visual field. Similarly, neural eye-movement neurons for the other two little eyes shift the relevant live connections. As Mary moves about, each neural eye may then learn to track what it sees, latching on to and following the visual feature it covers, always trying to maintain its original view of Mary. The neural signals from this original view of Mary at the center form a standard reference that the neural system remembers and always tries to match.

As our neural eyes follow Mary's movements, certain trends emerge. The three little eyes, for example, always tend to move together. If Mary moves to the right, all three neural eyes move to a similar extent. They exhibit correlations in their patterns of movements. Our neural system could learn these trends by feeding the output from the neural eye-movement neurons to what I call *correlation neurons*. Correlation neurons can capture the main ways in which their inputs tend to fire together. We need not go into the details of how this happens as it involves neural learning circuits and principles similar to those we already encountered for discrepancy neurons. As our neural eyes follow Mary around, the strength of synapses between neural eye-movement neurons and correlation neurons are gradually modified, so that the correlation neurons start to respond in a manner that reflects the way the neural eyes tend to move together. Through such learning, our neural system comes to expect this pattern of neural eye movements and any deviation from it would be registered as a discrepancy.

Having trained our neural system on Mary, what happens when we now look at John? Because our neural eyes have learned how to maintain a particular view of Mary, they attempt to do the same thing with John. A possible result is shown in figure 69. Our big eye and little eyes on the left and right are in same the positions as before, but the little eye in the middle has shifted its view to the right so as to keep the dark patch in sight. Having shifted its gaze, the middle neural eye now receives the same input as if it were looking at Mary. So, the visual signal entering each of our neural eyes is exactly the same as when our neural eyes were looking at Mary. We are looking at John but seeing him as we would normally see Mary. We are seeing John as Mary.

As far as the visual inputs to our neural eyes are concerned, there is no difference between looking at either person. But a difference is detected by the correlation neurons. These neurons detect a discrepancy when looking at John, because having captured the trend that all three little eyes move together (through neural learning based on Mary), they now find that the middle eye has moved differently from the others. The same discrepancy is always detected when looking at John, wherever he happens

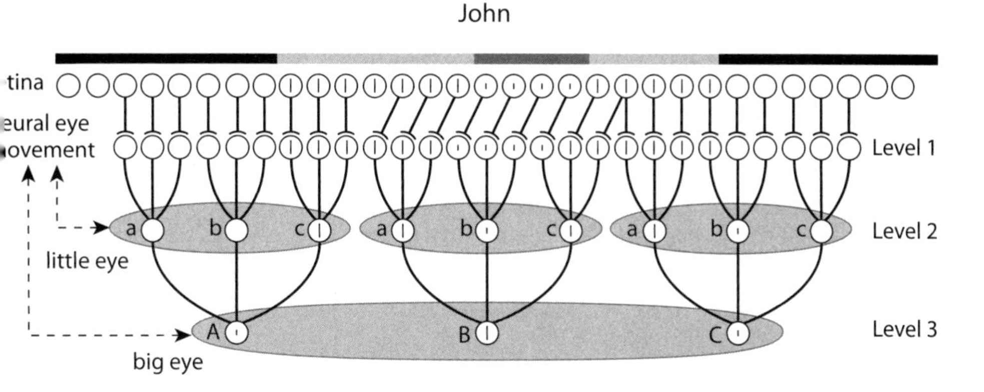

Figure 69. Looking at John after having trained neural eyes on Mary.

to be on the background. This distinctive shift in the middle little eye relative to the other neural eyes becomes a signature for John, a way of our neural eye system knowing that it is John and not Mary we are looking at.

Our neural system has accomplished what we set out to achieve. Our neural eyes tells us that John and Mary are the same kind of thing—we see John as Mary—but our correlation neurons tell us that there is also a difference between them; John and Mary elicit different firing patterns of correlation neurons. I have gone through this example in detail in order to highlight some general features of how we may learn to interpret what we see. Let's now look at some of these features in turn.

Seeing through Models

When looking at Mary, our neural system builds up an expectation, or a model, for what it sees. This involves the system learning what each neural eye normally detects and where the various neural eyes tend to go. I presented a simple one-dimensional example, but the same would apply to more complex objects we see around us. When you look at someone, their appearance may change in all sorts of ways. They may turn their head, move their mouth, or change their expression. These alterations are much more complicated than Mary moving from left to right. Nevertheless, as with Mary's movements, they involve elements of what you see moving in a correlated way. When a head turns, the eyes move together; when someone smiles, the left and right side of the mouth move in unison.

Your brain can somehow monitor these events, allowing you to develop expectations or models for how the various elements of the face tend to move together.

By forming a model in this way, you are making use of something normally taken for granted—physical continuity. You assume that the same person remains before you as you observe them. When continuously watching Mary for a period of time, you can be confident that someone else has not been substituted halfway through. This assumption allows you to build up a model for that person, a representation of how their different elements go together. If objects instead were continually flitting from one identity to the next, it would be very difficult for us to establish expectations or models. If an object were to switch from looking like an apple to a banana, or to a face, and then to a tree, it would be very difficult for us to form an impression of it. In practice, objects do not show such erratic behavior, so we are able to learn what are called invariant representations—models that allow us to perceive an unvarying identity when we are looking at something. We preserve this notion without reference to variations brought about by changes in pose or expression, just as we retained the same neural image of Mary as she moved around. In doing so, we may also learn how the various elements of our object behave in relation to each other. We learn the correlations involved when a face turns or smiles. The identity is maintained in spite of some changing elements.

Once we have established a model or expectation for an object, we may then be able to view related objects through that same model. We can come to see John through the model we formed when looking at Mary. This allows us to see one thing as a version of something else—we see John as Mary. But we are nevertheless aware that it is not exactly the same thing in front of us because of discrepancies with the model we had already built up. There is a double aspect to "seeing as." We have a common neural model through which we view related objects, but our neural system also registers discrepancies with the model, which allow us to know that one object is different from the other.

As we look at the new object for a while, the pattern of neural discrepancies we detect may themselves start to be learned. The signature for John that was initially viewed as a discrepancy becomes an expectation if we start to see John regularly. We can capture this expectation through learning circuits involving further correlation neurons, arriving at a broader model that includes both John and Mary. We might call this broader model a *person model*. When viewing John or Mary through the person model, each triggers a particular firing pattern of discrepancy and correlation neurons, enabling us to identify who it is. If we view a third person that does not look like either John or Mary, we would detect discrepancies with our person model, leading us to further identifications and elaborations. We are recurrently building correlation upon correlation to establish more general and elaborate models, learning from the pattern of discrepancies each time.

This process of building model upon model is part of the neural journey that each of us takes from birth. We all start with certain expectations (models), embodied in the neural connections and strengths with which we are born. We are also born with structured neural arrangements that allow these models to be elaborated and built upon through learning. For our neural scheme these structured arrangements were the various neural eyes and the way they behave together. We then modify our models through experience, learning particular correlations from the way the world changes as we view it, and then use this information to elaborate our models further. This is the principle of recurrence applied to neural models.

At the heart of this process of model building is learning through discrepancies and correlations. As we saw with the neural schemes for discrepancy neurons in the previous two chapters, this learning is based on our familiar double feedback loops between reinforcement and competition, fueled by a balance of variation and persistence. Based on a population of experiences (variation), particular neurons boost their own synaptic strengths (reinforcement) in a way that also brings about its own limitations (competition), and this leads to lasting changes in the strength of synapses (persistence). The same principles apply to learning schemes for correlation neurons, although we have not gone through these in detail. The feedback loops are more complicated because we now have a larger number of discrepancy and correlation neurons operating in combination, but the core principles are exactly as we have encountered before.

Learning at Many Levels

Another key feature of our neural scheme is that it involves interactions among multiple levels. We did not just have one neural eye; there were several, each working at a particular level. These neural eyes allowed us to go beyond what normal eyes can do. While a normal eye can only point in a single direction at any one time, our multiple neural eyes can point in several directions at once, allowing us to learn about many aspects of an object at the same time. This is what allowed us to see John as Mary—the little neural eye in the middle could look right while the other eyes looked straight ahead.

The various levels in our scheme do not work in isolation but in a highly cooperative manner. What the big neural eye sees depends on where the little neural eyes are looking, because the visual signal has to pass through the little neural eyes to reach the big neural eye. Conversely, what the little neural eyes see depends on the big neural eye, because movement of the big neural eye makes all their inputs shift along as well. This two-way relationship allows the eyes to assist each other. For example, one limitation of the little eyes is that individually they cannot detect large shifts because of their smaller visual field. A large shift might move the object out of their range. Such movements could most readily be captured by the big eye, which has a broader reach because it takes in the whole scene. By following a large movement, the big eye would effectively bring the object back into view for the little eyes, so that they can then detect any remaining smaller changes. Conversely, the big eye doesn't have the flexibility to follow details in the image, while the little eyes can track these and thus help deliver a more consistent picture to the big eye. Cooperation within and among multiple levels is critical for such systems.

In our simplified scheme, we had a retina with about thirty photoreceptors and three levels. But you could imagine a retina with many more photoreceptors and further levels. The human retina, for example, has on the order of one hundred million photoreceptors, so there are plenty of inputs that can go through many levels of integration. If we add an additional level to our scheme, each of our three little neural eyes would have its own three neural eyes that feed into it, so we would end up with thirteen (9 + 3 + 1) neural eyes looking at our object. With two further levels we would have 121 (81 + 27 + 9 + 3 + 1) neural eyes. That is, we could look in 121 different directions at once, each taking in aspects of an object at a particular scale.

Visual information is indeed dealt with at many levels in the human brain. Some of the levels are shown in figure 70. Signals from the photoreceptors of the retina first pass to neurons within the eye, called retinal ganglion cells. The outputs from the retinal ganglia go to a region of the

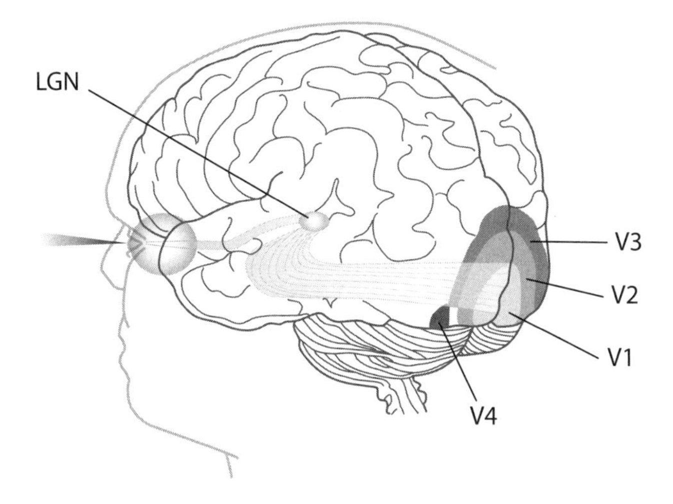

Figure 70. Some of the areas of the brain involved in visual processing.

brain called the lateral geniculate nucleus or LGN. The LGN sends outputs to the V1 region, which in turn sends outputs to a region called V2. This in turn is connected to other regions, such as V3 and V4. All of these regions, the retinal ganglia, the LGN, V1, V2, V3, and V4 are organized retinotopically—they all have a spatial pattern of sensitivities that mirrors that of the photoreceptors in the eye. Their visual information is also combined to various extents. The receptive fields for neurons in V4 tend to be larger than those in V2, which in turn tend to be larger than those of V1.

These brain areas are known to interact in various ways with each other and with other areas involved in vision. However, it is still unclear how these interactions allow particular visual interpretations to be learned. The scheme I presented of neural eyes moving around is one way of envisaging what might be taking place. Level 1 in my scheme relate to one of the visual areas of the brain. Neural eye movements would then correspond to shifting live inputs to this visual area. Key requirements are that the area corresponding to level 1 is organized retinotopically, and that there are further levels above this area that integrate its outputs and feed back information to modify its inputs.

Whether forming visual models involves the sort of scheme I presented with neural eyes or some other mechanism, having multiple levels that cooperate is likely to be critical. Imagine what it would be like if you looked at things through only one level. If you could only take in fine visual detail, you would acquire lots of information but have no idea of how to integrate it. Something like this happens in people suffering from *visual form agnosia*. When presented with an object they can see fine visual detail but cannot recognize the object itself. Conversely, if you could only take in visual information at the broadest level, you might obtain an overall picture, but with no detailed information you would have only a vague idea of what you see. It is only by seeing through multiple cooperating levels that we can arrive at particular interpretations and recognitions.

Multiple levels in vision do not mean that we need a fundamentally different mechanism for each level. Our little neural eyes, for example, operated in pretty much the same way as the big one. The little eyes involved abc neurons looking at one third of the visual field, while the big eyes involved ABC neurons observing the whole field. Their inputs entered from different windows, but the way the big and little eyes operated was the same. Similarly, we need not invoke special neural processes for each level of integration in our brain. The same basic elements may be recurrently used again and again, capturing relationships at a series of interconnected scales. This is similar to the way the same type of cellular interactions could be used to establish broad, or more detailed, features of a growing embryo (chapter 4, pages 93–97).

Top-down and Bottom-up

When we look at our surroundings, it is tempting to think that what we see is based on a chain of processing events. We receive information through our eyes, process it through a series of steps in our brain, and this tells us what we are looking at. This is sometimes called the bottom-up view of perception because it starts with the simple senses and ends up at the higher brain functions. By contrast, the top-down view is that we approach the world through particular models. These models are derived from us rather than from our senses, and allow us to see and interpret the sensory information we receive.

Our neural scheme illustrates how these two views—bottom-up and top-down—are not really alternatives but are woven together. As we view Mary, visual inputs travel from the photoreceptors of the retina to the various neural eyes at higher levels. The signals are travelling in a bottom-up direction. Changes detected in our neural eyes at levels 2 and

3 are then fed back to connections with level 1, influencing where the neural eyes are looking. We have top-down flow of information. This in turn changes the signals coming up from level 1 to the neural eyes at higher levels. We have continuous feedback among different levels, not a one-way flow of information.

Similarly, when we look at John, the model we have learned from observing Mary is used to direct our neural eyes. This is a top-down process because we see John through the model of Mary that we bring with us. But as we look at John for a while, our correlation neurons learn particular discrepancies, leading to a broader person model. This process involves bottom-up learning, because sensory inputs are now influencing the model. Again, there is continual feedback between models and the sensory information they detect. Interpretation comes neither completely from us, or from our sensory input, but from how the two continually play off each other through time.

In the previous chapter, we saw how our neural journey involves a continual two-way dialogue between what we do and what we experience. A similar two-way interaction applies to our interpretations. Just as we learn about how our physical actions relate to experiences, we learn how internal actions, symbolized as neural eye movements, relate to the objects they detect. Through continual feedback between internal neural activity and sensory information, we arrive at particular models or interpretations. These in turn guide additional interactions and experiences, and further elaborations of our models. Our journey through neural space is neither top-down nor bottom-up, but an inseparable mix of the two.

Competing Interpretations

Let's now revisit the problem of why we flip between alternative interpretations when viewing ambiguous pictures like the old woman and the young lady. Suppose that after having learned a person model for Mary and John, we are presented with an image that combines features of both, as shown at the bottom of figure 71. The left edge of the dark patch lines up with Mary but the right edge lines up with John. Our neural eyes and correlation neurons may be torn between two options—viewing it as Mary or as John. The little neural eye in the middle, for example, might be drawn to the right or to the center, or perhaps to a point in between. What happens in this situation? One possibility is that our neural system

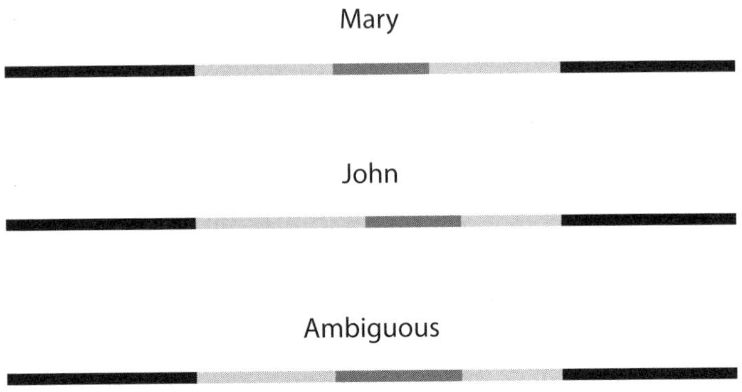

Mary

John

Ambiguous

Figure 71. One-dimensional John, Mary, and an ambiguous person.

sits on the fence—it sees the object neither as Mary nor John. Another option is that it goes for one of them, deciding that perhaps it is Mary. It may stick with its decision, or occasionally switch, sometimes viewing it as Mary, sometimes as John. Either way, some sort of decision is made rather than continually hovering in no-man's land.

It seems that our neural systems are structured so as to resolve situations, rather than sit on the fence. We tend to interpret ambiguous pictures either one way or another but usually do not dither in the middle. One way of achieving this resolution is through neural competition. Suppose that while learning our person model for Mary and John, we introduce competition between different interpretations. The correlation neurons that are triggered specifically by Mary tend to inhibit those triggered by John and vice versa. We end up with two mutually exclusive states—we tend either to see a Mary or a John, but not an entity in the middle. Other factors may then bias us toward one state or the other. If we have just seen Mary, we may be more likely to view an object as her because neurons involved in this interpretation were already firing. Or hearing Mary's voice may also tip the balance in her favor if we have previously learned to connect these sounds with Mary's appearance. Which way we go depends on all sorts of factors, but our neural system is organized to go one way or the other. We have a new level of competition—a competition between interpretations.

Why might our neural systems be structured to resolve matters rather than hover in intermediary states? The answer involves why we make interpretations in the first place. Suppose that Mary is a close friend while John is often unpleasant to you. Your expectations about the person in front of you then depends on whether you think it is Mary or John. If it is Mary you might predict a pleasant conversation, while if it is John you might anticipate a less enjoyable encounter. This in turn influences your actions—you are more likely to stay and talk with Mary while avoiding John if possible. At the end of the day, you have to decide what action to take, to engage or avoid, so it is important to work out who the person is. It therefore makes sense that our neural systems are structured to deliver particular interpretations rather than continually hover with no resolution. The fundamental reason we make interpretations is to help guide our actions, so it is important that we can resolve them.

To see how critical such resolutions are, imagine you went through life without being able to make decisive interpretations. You might not be sure whether to sit down on what looks like a chair, because in some respects it also resembles a table. You might wonder whether you should take a bite out of an apple, because it also looks somewhat like a tennis ball. Or you might think about opening a door but hesitate, because it also resembles a window. Life would be impossible if we did not continually resolve our interpretations one way or another. Without such resolutions we would be unable to act, trapped in a world of indecision. This problem is solved by having interpretations compete with each other. Through neural cross-inhibition, interpretations can become mutually exclusive, so we tend to categorize or separate our experiences out one way or the other. Such categorization helps us to act in particular ways, to decide to eat an apple because it is clearly different from a tennis ball, or go through a door because it is distinct from a window. Our categories are not written in stone—they can be modified through learning. Nevertheless, by forming some sort of categorization we are able to carve up our environment in an effective way. As we saw when considering biological development, we carve up the world by carving up ourselves. But rather than through inhibitory cell-cell interactions that define regions in an embryo, this self-carving now takes place by neurons inhibiting each other according to our experiences after birth.

Our interpretations depend on the same principles as those encountered throughout this book. The double feedback loop of reinforcement and competition underlies learning through discrepancies and correlations, as illustrated by our neural schemes for discrepancy neurons in the previous two chapters. These twin loops are fed by our population of experiences and actions, leading to persistent changes in synaptic strengths and numbers. The loops are also embedded in various forms of neural cooperation, competition, combinatorial richness, and recurrence. There is cooperation among multiple levels, symbolized by our neural eyes. These act in combination to yield a variety of models or interpretations for what we see. Interpretations also compete with each other so that we come to see an object one way or another. And all of this happens by recurrently shifting discrepancies and expectations. The complexity of our interpretations arises not from introducing new principles, but from further ramifications in the way our familiar principles operate.

A Question of Style

To show how the same principles can lead to quite elaborate interpretations, I want to continue building on our scheme of neural eyes and apply it to more complex images. The aim is to capture one of our more elusive interpretations, namely, an artist's style.

In our simplified example of one-dimensional people, we had three levels of neural eyes learning to follow particular aspects of a person, such as the dark patch in the middle. Now imagine extending this scheme to two dimensional images and multiple neural levels, with perhaps hundreds or thousands of neural eyes at the equivalent of level 2. As we view complex objects, such as a face, over a period of time our neural eyes learn to follow aspects of the face at various scales. A big neural eye may follow the face as a whole, while a smaller neural eye looking towards the middle of the face might follow the tip of the nose, or one looking further down might follow the chin. This notion is illustrated in figure 72, which shows where some of these smaller neural eyes may be looking when we observe portraits by Rembrandt (*left*) or Modigliani (*right*). I have indicated the positions looked at by neural eyes with spots placed over features such as the middle of the chin, the left mouth edge, and so on. Each spot is placed in a fixed relation to the others—white spot on the left eye, black on the right eye, and so on. I was able to position the spots in this way because I already know how to look and interpret two dimensional images of faces. I can readily find corresponding features, like the chin or

Figure 72. Positions looked at by seven neural eyes on a Rembrandt and Modigliani *Left*: detail from *Portrait of Maria Trip*, Rembrandt van Rijn, 1639. *Right*: detail from *Portrait of a Girl,* Amedeo Modigliani, 1917–18.

left eye, in both portraits. But you should imagine that these locations are found automatically by particular neural eyes because of the way they have been trained through exposure to many faces in the past. Indeed, I am only able to identify features like eyes and noses so effectively because my own brain has been exposed to such previous training.

If we look at numerous portraits in this way, there is an average position where each neural eye tends to go. For an indication of this average, look at figure 73. This image was generated by my placing positions for neural eyes on 179 portraits by artists such as Rembrandt, Modigliani, and Leonardo. As well as the seven positions illustrated in figure 72, further positions were also noted, corresponding to additional neural eyes. The average location of these neural eyes over all the portraits was then determined (only the average positions for our seven neural eyes are shown in figure 73, *left*). All of the portraits were then aligned and morphed according to these average positions, and superimposed to create an averaged portrait. This portrait represents an average based on a population of 179 portraits.

The averaged portrait represents an expectation, or model, for where our neural eyes tend to go. When we look at any individual portrait, the

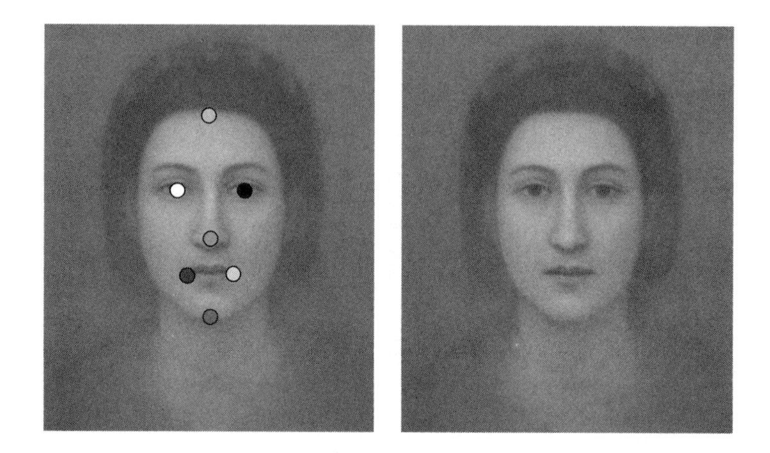

Figure 73. Averaged portrait, with average positions looked at by neural eyes superimposed (*left*). The 179 portraits used to construct the averages are by Rembrandt, Modigliani, Leonardo, Velázquez, Soutine, Giotto, Duccio, Freud, and a selection of Roman mummy portraits.

neural eyes do not go to precisely these average locations, but deviate to some extent. The neural eyes on the portraits by Rembrandt and Modigliani in figure 72, for example, are not in exactly the same positions as those on our averaged portrait. Each particular portrait therefore represents a deviation from our average, a discrepancy from our expectation. The discrepancies are not random, but show certain trends. In a portrait of someone turned to the left, many of our neural eyes shift together in the same direction. Or if the portrait is of someone smiling, neural eyes around the mouth move upward in a coordinated way. We can capture these trends with correlation neurons. Through reinforcement, competition, and cooperation, these neurons can distill the various trends—the ways in which our neural eyes tend to deviate from their average locations as we view the population of 179 portraits.

This process of neural correlation and distillation can be simulated with a computer. The details need not concern us here; all that matters is that the computer calculates the trends that might be captured by particular correlation neurons. The result for two of the correlation neurons is shown in figure 74 (plate 10). The middle picture for each of the two panels shows the averaged portrait, while the pictures on either side illustrate the trends captured by the correlation neuron. The neuron in the top panel has essentially learned to respond according to variation in the shape of the neck and face among the population of portraits. It fires at a high rate for portraits with a long neck, long nose, narrow head, and narrow lips, but at a low rate for portraits with the opposite features. I have called this neuron the *long-thin-head neuron*, naming it after what makes it fire at a high rate. I do not mean to imply that we have a long-thin-head neuron in our brain. The properties of the long-thin-head neuron are a consequence of the rather artificial task we have set: to find the trends in the way corresponding positions vary over a certain collection of 179 portraits. Although correlation neurons most likely exist in our brains, one with this particular response is unlikely to be among them. What the long-thin-head neuron illustrates is how trends may be distilled from a population of images through neural interactions, once we have a way of identifying corresponding positions (e.g., through neural eyes).

The correlation neuron in the lower panel of figure 74 responds in a different way. This neuron fires at a high rate for portraits with a long nose and neck not wearing a hat, but at a low rate for a portrait with the opposite features. I have called this the *hatless-long-head neuron*. In our collection of portraits it so happens that portraits of subjects with longer necks and noses (such as those by Modigliani) usually don't have hats. This correlation is captured by the hatless-long-head neuron.

When we view a particular portrait, our correlation neurons would fire at particular rates. A Modigliani portrait with a long thin head and no hat, for example, would make both the long-thin-head and hatless-long-head neurons fire at a high rate. I have shown the firing patterns elicited by some portraits by Rembrandt and Modigliani in figure 75. In this diagram, the firing rate of the long-thin-head neuron forms the horizontal axis, while the firing rate of the hatless-long-head neuron forms the vertical axis. The particular combination of firing levels elicited by each portrait can then be represented by the point where the firing levels intersect. These positions (calculated by the computer) are shown as light gray dots for Rembrandt portraits and dark gray for Modigliani portraits. Notice that Modigliani's paintings tend to be concentrated around the top right of the diagram. This is because most of them have long thin heads and no hats. They therefore elicit strong firing of both the long-thin-head neuron and the hatless-long-head neuron. Modigliani's *Portrait of Jeanne Hébuterne*, for example, strongly triggers both the long-thin-head neuron

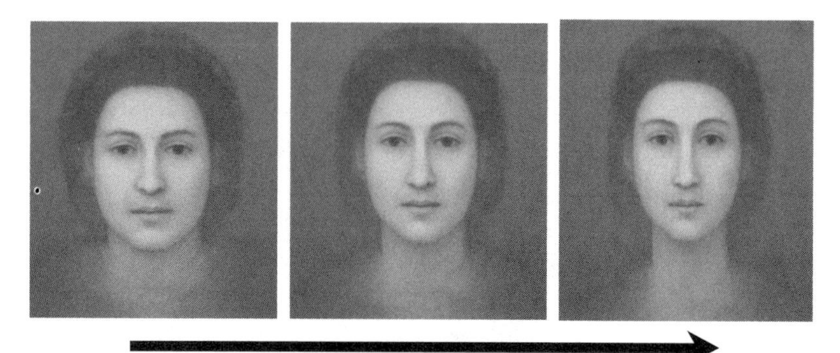

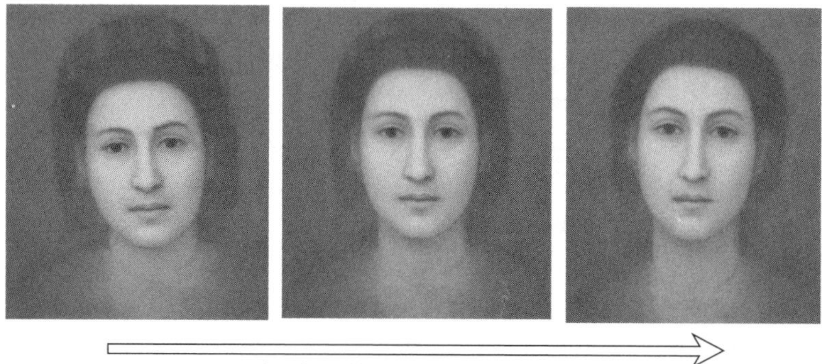

Figure 74. Two of the trends captured by correlation neurons from the collection of 179 portraits. See plate 10.

and the hatless-long-head neuron, placing it at the extreme top right. There are some exceptions, however. Modigliani's portrait of *Oscar Miestchaninoff*, for example, is located near the center rather than the top right. This is because not all of Modigliani's portraits have long thin heads. His portraits of men, in particular, tend to have much rounder faces and shorter necks, placing them among the Rembrandts.

By denoting Rembrandt's portraits with light gray spots and Modigliani's portraits with dark gray spots, I used information I had about the paintings—I knew which artist painted each. I knew this because the portraits were labeled with the artist's name. Similarly, when you view a portrait, as well as the visual input from the picture, you may have ad-

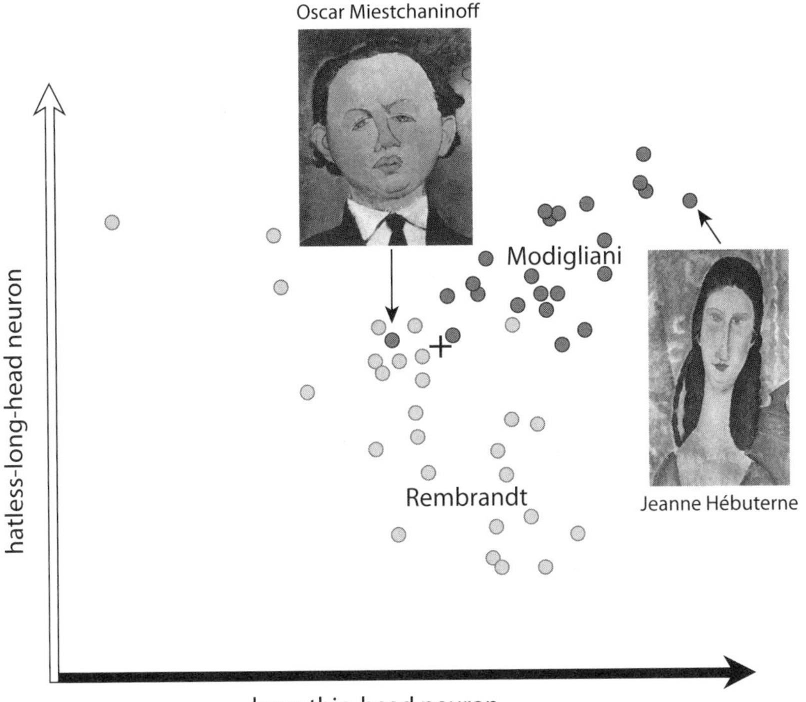

Figure 75. Firing patterns of correlation neurons elicited by portraits by Rembrandt (light gray spots) and Modigliani (dark gray spots). The cross in the center shows the position of the averaged portrait. Two individual portraits are shown. *Right*: *Portrait of Jeanne Hébuterne*, Amedeo Modigliani, 1919. *Top*: *Oscar Miestchaninoff*, Amedeo Modigliani, 1917.

ditional information as to the artist responsible. You may see a label with the artist's name, or someone may tell you the name of the artist responsible. In such situations, neurons involved in recognizing the painter's name would be firing when viewing a Modigliani portrait, in addition to the hatless-long-head and long-thin-head neurons firing. This correlation may be learned, allowing us to arrive at what we may call a Modigliani neuron. The Modigliani neuron captures the main trend for images which we know to be by Modigliani, illustrated by an arrow passing through the group of Modigliani portraits in figure 76.

The portrait image placed on the graph in figure 76 shows the deviation, relative to the averaged portrait, that corresponds to the Modigliani

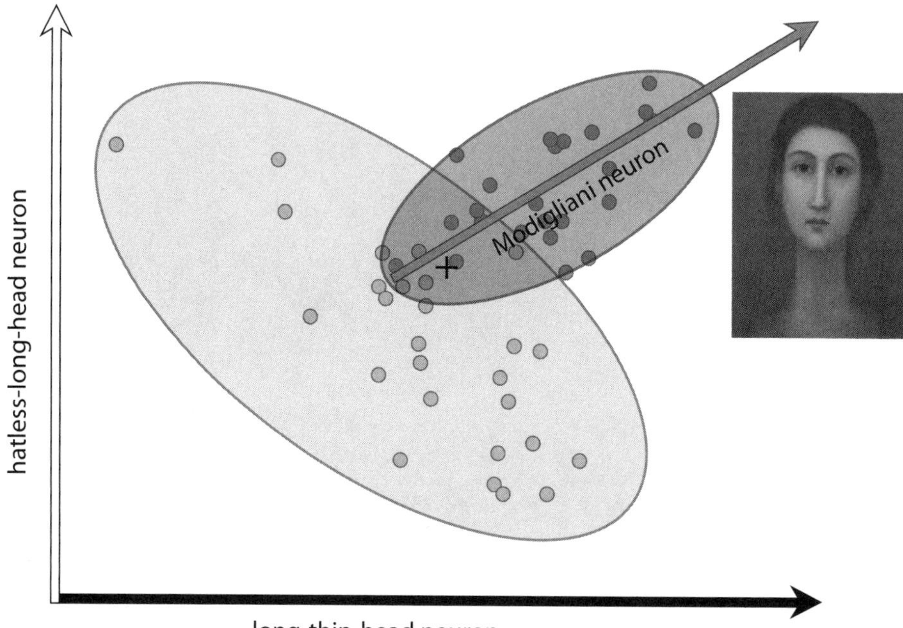

Figure 76. Trend captured by the Modigliani neuron.

neuron firing at a high rate. There are aspects of this image that are typical of many Modigliani portraits—a long thin head and no hat. Even higher firing levels of the Modigliani neuron corresponds to the image shown to the right of figure 77. In this case I have only shown the image outline, which looks rather like a caricature of Modigliani's style. The resemblance to a caricature is no accident—caricatures succeed because they are particularly effective in triggering neurons, like our Modigliani neuron, that are normally involved in recognizing the subject. Caricaturists are experts at manipulating our neurons.

We can now use the Modigliani neuron to help us determine whether a portrait we have never seen before is likely to be by Modigliani. If when viewing the portrait, our Modigliani neuron fires at a high level, then we would be reasonably confident that we are looking at a Modigliani rather than a Rembrandt. We have learned some elements of Modigliani's style. A connoisseur may learn to recognize a style in a similar way. By viewing

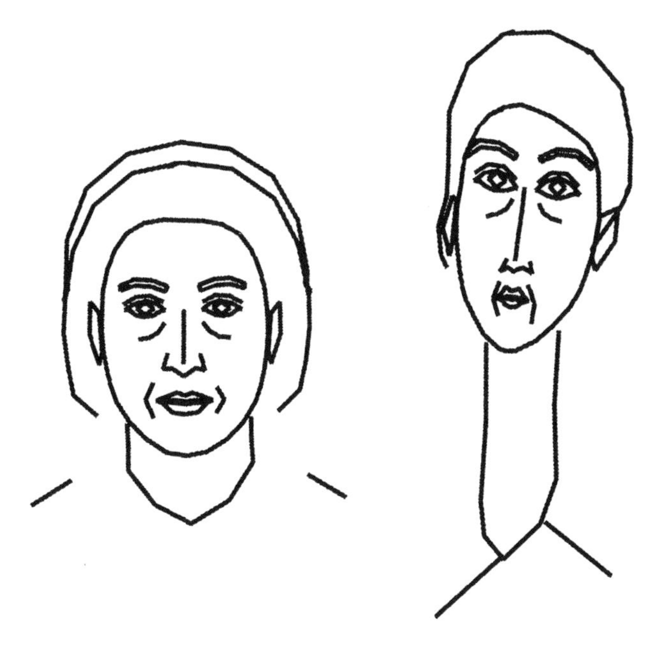

Figure 77. Outline corresponding to very high firing of the Modigliani neuron (*right*) compared to the outline of the averaged portrait (*left*).

many known examples, the connoisseur forms neural models, or expectations, for an artist's work. Once appropriate models have been established, an artist's work may trigger the relevant neurons in the connoisseur's brain, allowing them to identify the artist, even when there is no label. Connoisseurs never learn in a vacuum; they use prior knowledge about objects of known provenance to develop models for evaluating unknown objects encountered in the future.

The example I have given of extracting trends from a selection of portraits through neural interactions is highly artificial and simplified. I only described a few trends from a limited collection of portraits. I also only took account of a few neural eyes, and ignored aspects of the paintings like contrast or color. In practice, we have far more neural patterns available to learn from than I have indicated, allowing us to be much more effective at discerning style. That is why we can see that Modigliani's painting of *Oscar Miestchaninoff* is different from Rembrandt's portraits, even though it overlaps with Rembrandt's grouping in our graph. Nevertheless, the basic elements of stylistic recognition may be no different from what I described for the Modigliani neuron.

In arriving at the Modigliani neuron, I did not have to introduce any new fundamental principles, only further degrees of interaction in our neural scheme. The Modigliani neuron is no different from any other neuron; it acquires its properties purely through interactions with other neurons, rather than from some extra ingredient. The aspect of style captured by the Modigliani neuron is not a feature of the neuron alone, but results from the combined effect of a large number of the bottom-up and top-down neural interactions. Moreover, the output from the Modigliani neuron may itself feed back to influence these interactions. After a Modigliani neuron has been established through learning, seeing a woman with a long face and neck may trigger this neuron, reminding us of a Modigliani portrait. To some extent, we see the person as a Modigliani painting, while also being aware that we are seeing a person not a painting. Triggering the Modigliani neuron is not only an output, but feeds into our neural system and influences how we see things.

Although an artistic style may be hard for us to define, the underlying neural mechanisms need not be so special. They may simply reflect how our familiar creative recipe operates as multiple neural levels, models, and experiences interact with each other.

Creative Acts

Imagine Cézanne working on a canvas. After placing some initial patches of color, he may feel that some meet with his expectations, while others are not quite right or create surprising effects. He may then react to this mixture of expectations and discrepancies by adding more brush strokes, building the picture up, leading to further judgements and comparisons. The painting proceeds in this way as a critical dialogue between the Cézanne and what he sees in front of him. He may be conscious of some of this dialogue, but much of it may also be happening nonconsciously—he may instinctively feel compelled to add some color here or there, or feel that a brush stroke works well or is not quite right.

This creative dialogue involves a complex interplay between all the processes we have encountered—predictions, actions, and interpretations. Cézanne did not simply mix and place colors at random; based on

Plate 1. *Portrait of Ambroise Vollard*, Paul Cézanne, 1899.

Plate 2. *Portrait of Ambroise Vollard*, Pablo Picasso, 1909.

Plate 3. *The Rue Montorgueil, Paris, Celebration of June 30, 1878*, Claude Monet, 1878.

Plate 4. *The Garden at Les Lauves*, Paul Cézanne, 1906.

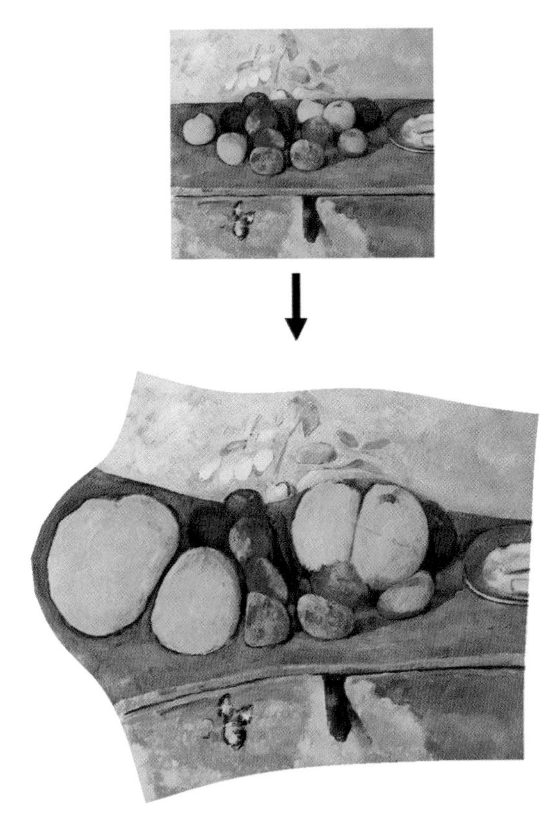

Plate 5. *Apples and Biscuits*, Paul Cézanne, c. 1880 and deformed version caused by increased growth of yellow regions.

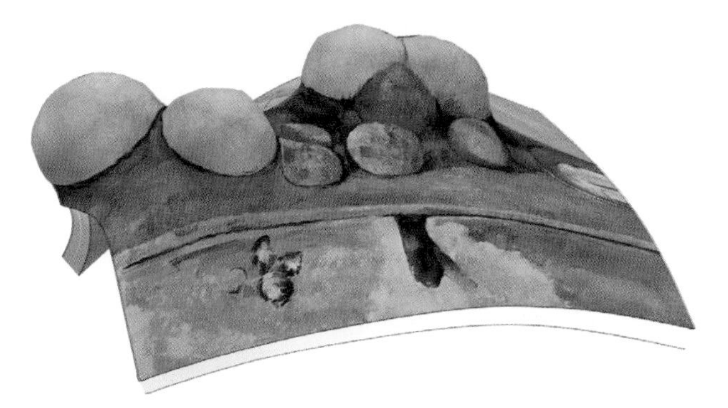

Plate 6. Deformed version of Cézanne's *Apples and Biscuits* when the canvas is allowed to buckle in three dimensions.

Plate 7. *The Librarian*, Giuseppe Arcimboldo, 1565.

Plate 8. "The Four Seasons": *Spring* (*top left*), *Summer* (*top right*), *Autumn* (*bottom left*), and *Winter* (*bottom right*), Giuseppe Arcimboldo, 1573.

Plate 9. Eye movements tracked for twenty seconds on *Still Life with Kettle*, Paul Cézanne, c. 1869.

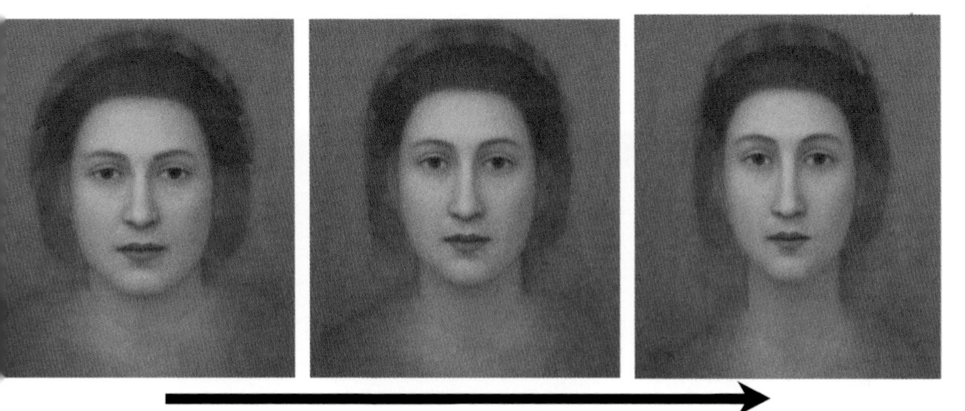

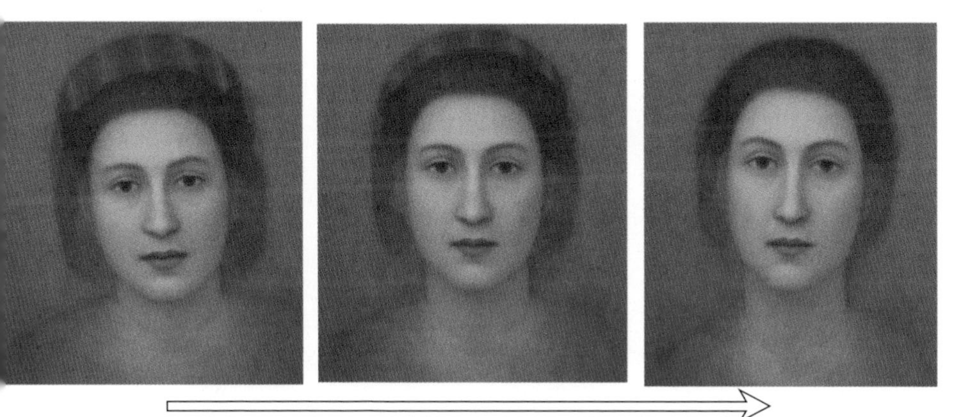

Plate 10. Two of the trends captured by correlation neurons from the collection of 179 portraits (see pp. 222–24).

Plate 11. *St. Jerome and the Lion in the Monastery*, Vittore Carpaccio, 1501–9.

Plate 12. *Hercules and the Hydra*, Antonio Pollaiolo, c. 1475.

Plate 13. *Top*: *Tobias and the Angel*, Andrea del Verrocchio and Leonardo da Vinci, 1470–80. *Bottom*: *Tobias and the Angel*, Antonio Pollaiolo and Piero Pollaiuolo, 1460.

Plate 14. *Still Life with Quince, Apples, and Pears*, Paul Cézanne, c. 1885–87.

Plate 15. *Apples and Pears*, Pierre-Auguste Renoir, c. 1885–90.

his experience as a painter he predicted what sort of effect they were likely to have and used this knowledge to guide his actions. He then interpreted the colors he placed on the canvas, exploring their relationships and comparing them to the subject he was painting. The matches and discrepancies he found then led to further predictions and actions, and so the process continued. The painting emerged through a neural dialogue that wove together predictions, actions, and interpretations at multiple levels. By painting a picture, Cézanne may also have increased his skills, modifying his neural models so that he approached a painting in a slightly different way next time round. The result is not only a painting, but the propulsion of Cézanne's brain through neural space.

Similar considerations apply to any creative act. When a poet composes a poem, there is a continual neural dialogue going on in their brain involving verbal sequences. Unlike painting, no overt physical actions are necessary—the poet need not write things down or speak aloud while composing. But as we saw with neural eyes, there is a fine line between physical and internal actions. A neural eye movement can have the same consequence on perception as physical movement of an eye, yet the first involves an internal adjustment of neural activity (of which connections are live), while the second involves an overt physical movement. The only difference between the poet and painter is that the poet's performance can be carried out purely internally, without necessarily involving physical actions and their sensory effects. This does not mean that the poet operates in a sensory vacuum. The language used by the poet was learned through prior interactions with others and with the environment. Even though a poem may be composed in a dark silent room, it builds on an extensive neural journey through which the poet has learned many relationships between sensory predictions, actions, and interpretations.

The same applies to the creativity of a scientist. The scientist's brain also toys with expectations and discrepancies, based on learning and experience. Through this dialogue, certain discrepancies may be resolved, leading to new models or ways of understanding. These may in turn lead to particular actions, such as the scientist carrying out an experiment to test a model. Experimental results may then feed back to influence further steps in the neural journey, leading to additional refinements of the model. As with painters and poets, much of the creativity of a scientist takes place nonconsciously, without the scientist being wholly aware of what is going on. A new explanation may seem to suddenly occur to a scientist, just as a poet may unexpectedly think of a lyrical line. Darwin records the pleasure he felt when he suddenly recognized how species divergence was related to the diversity of the environment: "I can remember the very spot in the road, whilst in my carriage, when to my joy the solution occurred to me." He had no doubt been worrying about this problem for a long time, with discrepancies circulating in his brain. This eventually led to a modification in his neural models that reduced certain discrepancies, resulting in the sudden realization that he felt in the carriage.

Creative acts of artists, poets, and scientists, or someone simply arranging flowers in a vase, can be viewed through the processes we have already encountered. They involve a recurring interplay between prediction, action, and interpretation. We have seen that each of these processes depends on learning through discrepancies and correlations. And these in turn revolve around our double feedback loop of reinforcement and competition, driven by a balance of population variation and persistence. Creative acts do not require any special new ingredients or principles, only further levels of cooperation, combinatorial interaction, and recurrence. They can be seen as a manifestation of life's creative recipe.

This may seem a somewhat impoverished view of creativity. How can something as wonderful as our own ability to create be reduced to a recipe of interactions that is fundamentally no different from that found in other living transformations? Surely human creativity is qualitatively different from a dog learning to salivate at the sound of a bell, or a monkey reaching to get an apple reward. I believe such a qualitative judgment is no more valid than saying that a frog is qualitatively different from the egg it came from, or that the swimming bacterium *E. coli* is qualitatively different from the first living organisms that emerged a few billion years ago. These differences that seem qualitative to us do not arise because of some mysterious new element; they are consequences of the way the ingredients of life's creative recipe operate and interact.

Ever since Darwin, scientists have tried to draw a clear line between creativity and transformations in the biological world. This approach was needed to overturn the misconception that living processes are accounted for by an external guiding hand or creator. But this separation has had the unfortunate effect of disconnecting human creativity from other living processes—it places human activity in a special category, separating us from our animal relatives. While this separation may be a source of comfort to some, I believe it is a misleading and impoverished view. To my mind, appreciating the formal similarities between our creative acts and other living processes gives us a richer perspective on all life and a more appropriate view of our place within it. It does not diminish the wonder of human creativity, but places it in a broader biological context.

So far, I have emphasized similarities between learning, evolution, and development—they can all be seen as manifestations of the same creative recipe. But there is another connection between them. Just as we saw that development is grounded in evolution (chapter 5), learning is embedded in both evolution and development. Our three instances of life's creative recipe are connected through history as well as form. I now want to look more closely at this historical relationship.

TEN

Framing Recipes

The story of Sinbad the Sailor begins on a narrow and busy street in Baghdad. One day, when a porter was walking along the street carrying a heavy load on his head, he decided to sit down to take a rest outside one of the houses. The porter heard singing and smelled delicious food coming from the house and its garden, making him exclaim how unfair it was that some people had so much pleasure for so little effort, while he suffered carrying heavy loads for small rewards. Hardly had he finished saying this, when a servant came out of the house and told the porter that his master wanted to speak with him. The master, Sinbad, spoke kindly with the porter and made room for him at the feast. Sinbad then went on to relate the story of his many adventures and journeys to the porter and the other guests.

This introduction to Sinbad the Sailor is known as a frame story. It sets the scene for the tales that Sinbad will tell, providing a story that frames other stories told within it. The frame story of the porter is itself framed by another story. Sinbad the Sailor is among the tales in *One Thousand and One Nights,* a book of many stories that a young woman Scheherazade told a king. Scheherazade has to keep the king entertained with a story every night to prevent him from executing her. The stories include that of the porter and Sinbad.

These stories told at various levels share some similarities. All of them are told to entertain a particular audience—*One Thousand and One Nights* is written to entertain its readers, Scheherazade entertains the king, and Sinbad entertains his guests. The stories are also rounded off at the end with satisfying conclusions. *One Thousand and One Nights* ends with the king marrying Scheherazade, Sinbad's storytelling ends with the porter realizing how Sinbad had really earned his keep, and Sinbad's tale ends with his final return to Baghdad. The various stories have a common form while also framing each other.

A similar type of relationship applies to the processes of evolution, development, and learning. Their common form is their basis on life's creative recipe. Their framing relationships involve their historical origins. Evolution frames development, because it is through evolution that the process of development arose. And both evolution and development frame learning, because it is through these prior transformations that learning became possible. We have already looked at the relationship between evolution and development (chapter 5). In this chapter I want to look at how evolution and development frame learning.

Development of Learning

Learning does not start with a blank slate, but is always based on a particular neural context. The TD-learning described in chapter 7, for example, requires particular connections among neurons involved in detecting rewards, discrepancies, expectations, and sensory inputs. Similarly, calibration and the ability to form particular interpretations both depend on preexisting neural arrangements. These initial connections and relationships trace back to the process of development. Through the development of embryos, animals are born with highly structured brains, bodies, and sensory systems needed for learning.

The transition between development and learning is not sharp, however. Unlike the storytelling of Sinbad, which begins with a particular sentence, the transition between development and learning is much more blurred. Firing of neurons does not begin with birth but is already happening during embryonic development. And development does not stop with birth but continues as the newborn grows and matures. A good illustration of the subtle interplay between development and learning comes from some studies on vision.

In the previous chapter, we encountered a region at the back of the brain involved in vision, called the primary visual cortex or V1. If you shine a spot of light onto a screen in front of an anaesthetized cat or monkey, a small region of V1 shows strong electrical activity. As you move the light from one place to another in the visual field to stimulate a different part of the retina, the region of electrical activity in V1 also moves. It is as if there is a map in V1 that is connected to a corresponding map on the retina. The V1 area of the brain is said to be organized as a retinotopic map.

Animals usually have two eyes, so there are two potential retinotopic maps in V1, one for each eye. How are these maps integrated? In the 1960s, David Hubel and Torsten Wiesel working at the Harvard Medical School in Boston, started to address this question by make detailed recordings from the visual cortex of cats and monkeys. Like us, the left and right eyes of these animals have overlapping visual fields. Hubel and Wiesel found that when a short bar of light is shown at one location on a screen, first to one eye and then the other, a similar region of V1 is activated in both cases. Moving the bar to a new location shifts the region activated in V1 for both eyes in parallel. So the connections from both eyes are broadly aligned with each other in V1.

However, they also found a finer scale pattern. Within a given region of V1, they observed that some neurons mainly respond to light falling on the left eye, whereas other neurons react mostly to light falling on the right eye. The inputs from the two eyes were to some extent kept separate, even though they mapped to the same region of V1. They eventually realized that there were two tightly interwoven maps in V1, a left version and a right version. The relationship between these maps for V1 of the macaque monkey is shown in figure 78. The black regions show where one eye dominates, say the right eye, while the white regions show where the other eye dominates, the left eye. It is as if there are two copies of the visual map, one for each eye, that are slightly staggered relative to each other. The result is an alternating patchwork of left and right inputs, known as the ocular dominance pattern. How does this pattern become established?

Hubel and Wiesel decided to look at what would happen if animals were deprived of sight in one or both eyes as they grew up. Kittens first open their eyes ten to twelve days after birth. To deprive a cat of sight in one eye, say the left eye, Hubel and Wiesel surgically sewed the left eyelids together in a newly born kitten, depriving that eye of any visual input. When the left eye was surgically opened a few months later, they found that the kitten had become irreversibly blind in that eye. Instead of responding to both eyes, V1 only had strong connections to the right eye,

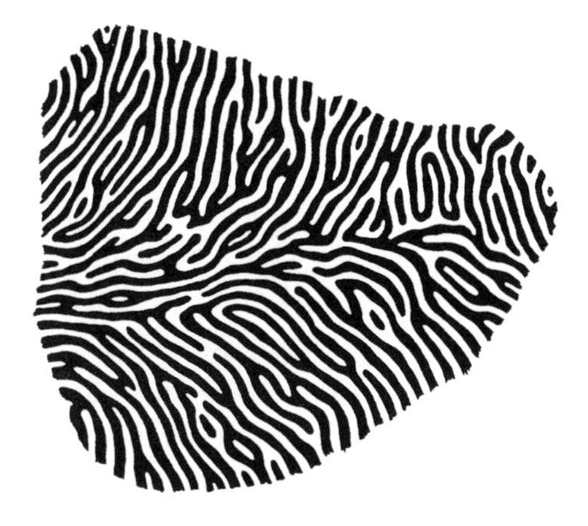

Figure 78. Ocular dominance patterns in primary visual cortex of macaque monkey.

the one that had remained open. At first they thought that the weakened synaptic connections for the closed eye were simply a consequence of disuse. To test this idea, they repeated the experiment but now surgically closed both eyes after birth, expecting that vision in both left and right eyes would be severely impaired. Surprisingly, they found that the connections were not that bad, as if closing a second eye had actually helped the first. It was not simply disuse causing blindness after temporary sight deprivation, but instead there was some sort of interaction between the two eyes.

The interaction is best described by returning to the ocular dominance pattern of figure 78. When one eye is surgically closed, rather than having an ocular dominance pattern with alternating bands of equal width, the bands from the active eye become much wider. One set of bands, say the black bands in figure 78, spreads out leaving only narrow strips of white corresponding to the closed eye. This means that the cat has lots of neural connections to the open eye and relatively few to the other. By closing both eyes, neither eye takes over, so you end up with a banding pattern with similar widths for the left and right eyes again, as with a normal cat. It seems that the width of the bands depends on a competition between the eyes. If one eye is more active, then it takes over the space that would normally form connections with the other eye. If both eyes are equally active or equally inactive, then connections are shared more equally.

Later computational studies showed that the experimental observations of Hubel and Wiesel could be accounted for by mechanisms similar to those invoked by Alan Turing to explain spotty or stripy patterns (chapter 3). In the case of ocular dominance, however, reinforcement and competition do not just involve local molecular signaling, but the extent to which neurons fire. For a given region of the visual field, nearby inputs from the same eye tend to fire in a more highly correlated way than those from different eyes. This is because each eye views objects from a slightly different angle and so there are slight discrepancies between what each eye sees. Computer simulations show that if greater correlation in firing leads to reinforcement of synapses, and if there is also inhibition among nearby neurons, then the alternating pattern of connections to the left and right eyes can be accounted for. When one eye is surgically closed, only the connections from the active eye are reinforced while the inactive ones are weakened and eliminated. This accounts for the observed spread of the ocular dominance bands for the active eye and their decline for the other.

The banding pattern does not depend solely on visual experience, however. As mentioned earlier, if both eyes are kept closed after birth, an ocular dominance pattern still forms in the visual cortex, although it begins to decay after about four weeks. This inborn tendency to form a banding pattern depends on the process of development, and may involve spontaneous neural activity in certain regions of the brain. Rather than there being a sharp line between development and neural changes brought about by sensory experiences, there is close interplay between the two. Development leads to particular patterns of neural organization; experiences then modulate and build upon these patterns, allowing relationships between neurons to become related more precisely to what an animal experiences.

This blurred transition between development and learning is possible because the two processes share many of the same molecular and cellular mechanisms. As we saw in chapter 6, short- and long-term changes in synaptic strengths involve molecular signaling, growth or demise of synapses, and switching on or off of particular genes. The molecular components of learning are no different from those involved in development. With learning, however, they are put together in a particular way that allows interactions with the environment to be captured through modification of neuronal pathways. Through this process of modification, neural journeys that began in the womb are driven in new directions, according to what is learned through actions and experiences.

But neural journeys are not only framed by development; they are also framed by the larger story of evolution. To understand this next level of framing, we have to look at why we learn in the first place.

Basic Instincts

The life of the fifth-century monk St Jerome has often been chosen as a subject for religious paintings. Episodes from Jerome's life have been depicted by artists including Dürer, Bosch, Leonardo, Caravaggio, and Rubens. But perhaps the most endearing portrayal is Vittore Carpaccio's depiction of Jerome and the lion (fig. 79, plate 11). According to legend, a lion limped into Jerome's monastery one evening, causing all the monks to scatter. Nevertheless, Jerome stood firm, realizing that the lion was suffering from a thorn stuck in his paw. Jerome removed the thorn and the grateful lion became his trusted companion, later helping to guard Jerome's donkey from danger.

The story of St Jerome is a play on instincts. Our natural tendency is to flee from a lion, yet Jerome overcomes this instinct by standing firm and helping the animal. And from the lion's viewpoint, the natural instinct is to attack humans. Yet the lion learns to curb this aggression as he is befriended by Jerome. Instincts are often presented this way, as something negative that are best overcome. But instincts in the broadest sense lie at the heart of all of our actions, including those most noble. This is because even when we think we have learned to overcome our instincts, that learning itself is grounded in our instinctive reactions. To see why, we must look at the evolutionary origins of learning.

Darwinian evolution is all about reproductive success. If particular responses to environmental change lead to increased survival and reproduction, these responses then tend to spread in a population through the process of natural selection. This accounts for the prevalence of instinctive responses such as the bacterium *Escherichia coli* swimming toward sugar, plants flowering at certain times of year, or slugs retracting in response to being touched. All of these responses can increase reproduc-

tive success because they have been tried and tested over many generations. But what is the evolutionary advantage of learning to respond to challenges that may never have been encountered before? Why should natural selection favor a dog learning to respond to the sound of a bell, or a monkey responding to the sound of an opening door?

From an evolutionary point of view, it would be most advantageous if organisms could learn to modify their actions according to what is likely to increase their reproductive success. The problem, though, is that by the time you are able to judge your ability to survive and reproduce, it is often too late to do anything about it. If, for example, you could have several experiences of being eaten by a lion, you might learn that avoiding lions is a good way of staying alive. But of course such learning is not possible because once you have been eaten by a lion, it is the end of the story. Similarly, if you wanted to learn which of several partners had the best reproductive potential, you might try producing and rearing children with each of them. But by the time you can predict which partner is best, you and your potential partners would probably be past your reproductive sell-by dates. Survival and reproduction rates are intrinsically difficult to learn about and act upon because they require many generations and individuals for effective measurement.

If learning within a lifetime is to be valuable for Darwinian evolution, it has to establish predictions about reproductive success more indirectly. The way it achieves this is by being connected to instinctive responses. Suppose, for example, that you have an inborn fear of large animals with sharp teeth, an instinctive response that could have been built up over many generations through natural selection. When you see a lion bare its teeth, you are instinctively afraid and run away. Now if you can learn about factors that help predict a lion is about to bare its teeth, such as seeing a shaggy mane or hearing a loud roar, you might run away without waiting around to see what the lion does. This action would help you survive. Similarly, your survival chances would increase if you could learn that lions are common in certain areas, allowing you to avoid these areas or be wary when you enter them.

Such learning works to your advantage because the instinctive fear has been put in place by natural selection. If you were born with a fear of apples, you would learn to become afraid of features that predict their appearance, like the sight of apple trees. But this would not help you survive and reproduce because being afraid of apples has no selective advantage to begin with. It is through connections with instincts that have been previously honed by natural selection that learning gains its evolutionary advantage.

This relationship between learning and instinct can be quite subtle. Infant rhesus monkeys are not at first afraid of snakes, but they soon learn to be if they see their mother frightened by a snake. This would seem to suggest that the fear of snakes is not instinctive but learned. Psychologists Michael Cook at the University of Wisconsin and Susan Mineka at Northwestern University explored this further by showing infant monkeys movie clips of adults looking fearful, followed each time by clips of various objects. Some of these objects were innocuous, like flowers or a toy rabbit, while others were more relevant to what is threatening in nature, like toy snakes or crocodiles. When these movies were shown to young monkeys, the infants acquired a fear of potentially dangerous objects, like snakes, but not flowers or rabbits. Evidently the infants already have a neural framework that predisposes them to become fearful of animals that presented hazards in the species' evolutionary past.

I have given examples of harmful aspects of the environment that are best avoided. But similar principles apply to desirable aspects of our surroundings. We may instinctively prefer the taste of particular types of food, like apples. If we learn that other features, such as color and smell, are good predictors of tasty apples, then we may use this information to improve our ability to obtain them. Similarly, we might be instinctively attracted to certain features of the opposite sex, such as their body shape. If we can learn other features that help predict such shapes, such as the type of clothes being worn or a way of walking, then these may help us find a possible partner.

Helping others is also a valuable instinct with an evolutionary basis. The action of assisting your offspring or relatives can be favored by natural selection, because it may increase the reproductive success of shared genes. Helping others can also pay through the return of favors. In the legend, St Jerome benefits from helping the lion by gaining a trusted companion who guards his donkey. This does not mean that Jerome was thinking of his own benefit when he helped the lion, only that the overall outcome of helping another can turn out to be advantageous. From an evolutionary point of view, the desire to help others may be thought of as an instinct like any other that can sometimes be favored by natural selection. This means that it can be advantageous to learn to recognize relatives or those who have helped us in the past.

All of these examples can be summarized by saying that we are born with a set of values—experiences that we consider as rewards or punishments, and that thus elicit actions of engagement or avoidance. These inborn values were put in place through natural selection. The initial values are built into the pattern of neural connections we are born with, and arise through development. Learning then allows these connections to be modified through experience, so that we increase our chances of obtaining the desirable and avoiding the undesirable.

Our initial values are only the start of the story. As we have seen, processes such as TD-learning lead to changed expectations and the introduction of new values (chapter 7). For humans, money is a predictor of being able to obtain food and other desirables, and so becomes valuable in its own right. Sinbad gets excited when he finds diamonds during one of his journeys because of the value he attributes to them. Values may also compete with each other and change according to conditions. Suppose you are torn between eating apples from a tree and running away because a snake is nearby. Your decision may be influenced by factors such as how much food you have recently eaten. If you are very hungry, the relative value of the apples is effectively raised and you are more likely to stay. Values may also change through development. Interest in sexual partners increases during puberty because of hormonal changes in the body. There is a complex interplay between conditions, development, and learning, which leads to values changing in all sorts of ways during our neural journeys.

Nevertheless, all these changes are built upon a foundation of some initial values that have been honed by natural selection, for that foundation provides the evolutionary rationale for our ability to learn. The chances of survival and reproduction are increased by learning to meet instinctive values, such as the desirability of eating when hungry and the undesirability of pain. If an animal was born with the opposite set of values, desiring pain and avoiding food when hungry, learning would actually be disadvantageous. The animal would learn how to avoid food and expose itself to pain more effectively, quickly starving itself or bleeding to death.

Animals with such behavior would be selected against. The values we are born with have been put in place through natural selection and provide the foundations for learning.

Flexibility versus Directness

We have seen how learning can be advantageous from an evolutionary point of view. But there are also some drawbacks. First, learning is only indirectly linked to reproductive success. This means that you can behave in ways that do not increase reproductive success, such as deciding not to have children because it costs too much. Secondly, learning is a far more elaborate process than instinctive responses, requiring more complex neural pathways and interactions. Why did learning evolve given that simpler and more direct instinctive responses already existed?

The answer has to do with trade-offs between flexibility and directness. One of the limitations of evolution by natural selection is that it can only capture trends in the environment that persist over multiple generations. But many features of our surroundings change over much shorter timescales, often ways that are hard to predict. If all our actions were completely inborn and took no account of this environmental variation, then many adaptive behaviors would be hard to achieve. Imagine, for example, that a Venus fly trap had to predict, without information from its surroundings, when to close its leaf on an insect. It would need a complex model of the world that would tell it precisely when an insect was going to land, a prediction that is almost impossible to make. Rather than prescribing all actions that are likely to increase reproductive success, natural selection has favored more indirect approaches.

A first step toward indirectness is to employ instinctive responses, as these are less direct from an evolutionary perspective than having all actions preordained from birth. The Venus fly trap adjusts its actions according to the environment: its leaves are triggered to close by insect movements. Instinctive responses like this provide a basic means of trading directness for flexibility. With responses rather than actions being inborn, we can act more flexibly according to changing conditions. This strategy underlies a whole range of actions, such as the bacterium *Escherichia coli* producing lactose-digesting enzymes when surrounded by milk, or swimming toward regions rich in sugars. These particular actions are not specified in advance but depend on what the bacterium encounters in its environment.

Instinctive reactions also have their limitations, however. These are particularly evident when it comes to moving about in the macroscopic world. A newborn, multicellular animal experiences all sorts of complex events brought about by its own movements. As it turns its head, its visual inputs change in complicated ways according to what is around it. As it begins to walk, it encounters many obstacles in its path, some of which are best avoided and others that are more beneficial. And to obtain food it needs to know where to find it and how to negotiate its complex surroundings to get there. Each individual encounters its own set of experiences, according to its particular settings and movements. Sorting out all of this complexity with only instinctive reactions would be extremely difficult. The animal would have to be born with an instinctive knowledge of all the types of objects it will encounter, how they might look from each angle, and how to adjust its movements to each in particular settings.

Rather than prescribing all of these behaviors through instinctive responses, another approach is to capture some of the complexity through learning. This allows the animal to tailor its bodily actions to its surroundings much more flexibly. The animal need not have inborn knowledge of exactly how to respond to obstacles and events, but instead it gradually modifies its actions according to what tends to bring about desirable or undesirable outcomes. It learns to move and carve up the world in particular ways, according to its experiences and values. The learned actions are likely to promote reproductive success because the values are themselves rooted in instinctive responses, honed by natural selection. Although more indirect than instinctive responses, learning provides greater flexibility for coping with the complexities and contingencies of what the animal encounters. Having such flexibility can increase reproductive success and has therefore been favored by natural selection. This is why our ability to learn and our patterns of learning have evolved.

Like the tales told by Sinbad, learning is framed by two larger stories—development and evolution. But there is also a further story to be told. Our three nested tales of evolution, development, and learning frame yet another tale. This story is about how humans have come together to bring about another type of transformation: cultural change.

ELEVEN
The Crucible of Culture

Societies of today are very different from those of ten thousand years ago. This is not because of major differences in biological makeup—if you had been born then, you would probably fit in perfectly well. Rather, it is because of dramatic changes in our culture and outlook. If you could travel back in time as an adult to meet your ancestors living ten thousand years ago you would encounter innumerable cultural contrasts. Your ancestors would immediately be struck by the strange clothes you wear, from your shoes to the shirt that is tailored to your body. If you attempted to greet them by shaking their hand they would probably look alarmed and wonder what you were about to do. As you got to know them better, perhaps learning a little of their language, you might try to convey our lifestyle to them. You might speak of how we live in houses with transparent sheets of material that let the light in, of how we eat our food with pronged instruments, and of how we often do not catch our meal but exchange it for pieces of paper. Or you might describe how we can move around very fast in boxes on wheels or fly in the air inside winged machines, and how we can talk and send messages to people that are very far away even when we can't see them. You might also try to explain that we all live on a large spherical mass that rotates, and how the seasons change as our sphere orbits the sun. After listening with incredulity perhaps your ancestors would decide to show you some paintings in their caves. You might then try to explain how it is possible to portray things in other ways, with the use of perspective or impressionistic splashes of color. If they sing to entertain you, you might pull out your media player and watch their quizzical reaction as they listen to a pop song, some jazz, or a symphony.

All of these contrasts in clothes, manners, living styles, economics, transport, communication, science, and art reflect how human societies have transformed over the centuries. There are many ways of describing how these cultural changes are brought about. We could emphasize the importance of power struggles, the forces of commerce, the leadership of charismatic individuals, the roles of innovation and imitation, and the vital part that language or means of transport and communication play. I want to look at the problem from a different perspective.

We have seen how we can arrive at some core principles by stepping back and looking at what is common to a range of living transformations—evolution, development, and learning. I now want to bring cultural change into the fold. Some of the principles we have covered have already been proposed to play a fundamental part in cultural transformations. The historians J. R. and William McNeill, for example, have emphasized how competition and cooperation have been critical for driving human cultural change. However, the full collection of ingredients and interactions in life's creative recipe has not previously been used as a way of considering cultural change. This is because there are so many elements of human behavior to choose from that the set of seven principles and their interactions does not come most obviously to mind. It is only by taking a broader perspective that encompasses all living transformations that this view of cultural change starts to crystallize. When it does, I believe we not only gain a more fundamental understanding of how cultural change is brought about, but also a clearer view of how it is related to our biological past.

To convey this viewpoint, it helps to give a particular cultural example. There are many that could be chosen, from war and fashion, to science and art. I have chosen an episode from Leonardo da Vinci's life because I want to show how even such an iconic example of cultural achievement can be viewed through the principles we have covered.

The Apprentice

In the middle of the fifteenth century, the goldsmiths Andrea Verrocchio and Antonio Pollaiuolo were fighting it out in Florence. At that time Florence was a thriving city, famous for its weaving, and home to the powerful

Medici banking family. Goldsmiths were therefore very much in demand by prosperous citizens for the production of ornaments and jewellery. Verrocchio and Pollaiuolo competed for the best commissions in this lucrative setting. But their ambitions did not stop with being top goldsmiths. Both wanted to increase their income and reputation further by branching out in new directions. Verrocchio decided to take up sculpture while Pollaiuolo trained as a painter. Pollaiuolo's first major painting was a series of three large canvases depicting the *Labours of Hercules*, completed in about 1460 for the Medici palace. This was a large-scale work, requiring assistance from Antonio's younger brother, Piero Pollaiuolo. Although the original canvases have been lost, a small version of *Hercules and the Hydra*, by Antonio Pollaiuolo (fig. 80, plate 12) reveals what they may have looked like. The paintings of the Pollaiuolo brothers were considered a great success and they won further lucrative commissions, becoming renowned Florentine painters.

Andrea Verrocchio was also successful in his new pursuit of sculpture, but perhaps spurred on by his rival's success, he also decided to take up painting in the late 1460s. Lacking a brother to assist him, Verrocchio enlisted the help of some assistants and apprentices. Over the years, his trained assistants included painters, such as Sandro Botticelli, Domenico Ghirlandaio, and Pietro Perugino, all of whom became famous artists in their own right. Two of the apprentices were Lorenzo di Credi and, most famously, Leonardo da Vinci. Leonardo was born in 1452 in the hill town of Vinci, twenty miles from Florence. He was the illegitimate son of a lawyer Ser Piero da Vinci and a country girl Caterina, and was brought up in his paternal grandfather's house. Leonardo showed a talent for drawing at an early age, so his father decided to place him as an apprentice with Verrocchio when Leonardo was still a teenager. This was around the time that Verrocchio was taking up painting.

The first painting to come out of Verrocchio's workshop in which Leonardo's hand can be reliably identified is a depiction of *Tobias and the Angel* (fig. 81, *left*; plate 13). It is believed that Leonardo contributed to features like the dog and the figure of Tobias. The Pollaiuolo brothers had also painted the same subject several years earlier (fig. 81, *right*; plate 13). The similarities in composition and details, such as the positions of the left arms and the dog at the angel's feet, show that the painting by Verrocchio and Leonardo was greatly influenced by the Pollaiuolos' work. By

Figure 80. *Hercules and the Hydra*, Antonio Pollaiolo, c. 1475. See plate 12.

Figure 81. *Left*: *Tobias and the Angel*, Andrea del Verrocchio and Leonardo da Vinci, 1470–80. *Right*: *Tobias and the Angel*, Antonio Pollaiuolo and Piero Pollaiuolo, 1460. See plate 13.

making such as direct reference to his predecessors' painting, Verrocchio was not performing an act of homage. Rather, he was using the comparison to highlight his superior skills in rendering stronger sculptural forms and a more intimate composition. It was one rival trying to show how he could surpass the other.

Let's now look at this story through the principles of life's creative recipe. We begin with the principle of population variation.

Fruitful Populations

Leonardo and Verrocchio were only able to achieve what they did because they belonged to a particular social setting, a population of individuals. In Leonardo's time, about fifty thousand people lived in Florence. This population and its nearby towns produced several great artists of the Renaissance, including Leonardo, Michelangelo, and Botticelli. It is unlikely that there was a special set of artistic genes that happened to then be circulating around Florence. Rather, the Florentine population of the fifteenth century represents a particular social setting that allowed the talent of individuals to flourish in particular ways. There are perhaps thousands of people alive today with a genetic potential equivalent to Leonardo's or Michelangelo's (today's world population is about one hundred thousand times greater than that of fifteenth century Florence). These modern equivalents do not produce masterpieces like the *Mona Lisa* or *God Creating Adam* because they were born and brought up in a very different setting. As we have seen before, every population assumes a particular context, be it a game of roulette, a forest of apple trees, a cup of tea molecules, or an individual's brain. For Leonardo's story, the context was Florence and its environs in the fifteenth century.

The population of Florence was not made up of identical individuals. Everyone was different because of variation in birth and circumstance. Leonardo, Michelangelo, and Botticelli each had a distinctive genetic makeup, so each started their life from a different position in neural space. The neural journeys they then took also depended on the particular encounters and experiences they had during childhood and beyond. Their interactions with relatives, fellow artists, and other citizens influenced particular interests and aspirations for each individual. These experiences and interactions included chance events, like Leonardo going to work in Verrocchio's studio when Verrocchio had just started to take up painting. As the art historian David Brown wrote: "Had Verrocchio not taken up the brush at this time, Leonardo might conceivably never have become a painter." The Florentine population was diverse because of the complex ways in which genetically different people interacted with each other and their surroundings.

Human diversity opens up many possibilities. If everyone was genetically identical and exposed to exactly the same experiences, then exchanges among individuals would be far less interesting and fruitful. But because everyone is different, they can continually learn and benefit from each other's knowledge and abilities. These exchanges can also influence each person's creative activities. As we saw in chapter 9, creativity is built into the way humans learn from and interact with the world. Exchanges among diverse people can result in a broad range of creative acts continually cropping up. Exceptional creative achievements, such as a particular painting or idea, may then bubble up in the population, becoming more noticeable than others. This may lead to the identification of particular individuals as being outstandingly creative. But such individuals never operate in complete isolation. Cultural advances depend on populations of diverse, creative individuals interacting with each other. As with other transformations, the principle of population variation provides a critical ingredient for cultural change.

I have given the population of Florence as an example, but of course this city did not function in isolation. The people of Florence could travel to other cities. Leonardo spent much of his later life in Milan, and Michelangelo traveled to Rome to paint the Sistine Chapel. Travel further afield was also possible, although it could take many months. The explorer Amerigo Vespucci, who was born in Florence two years after Leonardo, made several expeditions to America, which is named after him. Instead of isolated populations, the world comprised many interconnected ones. Population variation also applies to this broader setting that includes the diverse, interacting populations of the globe. While some advances were being made in Florence, others could be made in Samarkand or Beijing. And because of worldwide communication, these diverse populations could also interact and influence each other. Leonardo's oil paints had been invented a few years earlier by the Flemish artist Jan van Eyck, and the paper Leonardo wrote on was an invention that traces back to China many centuries earlier. The Florentine population did not operate in isolation; it was part of a broader population that spanned the globe, totaling about 350 million people in the fifteenth century. Without this broader backdrop of variation, the Renaissance in Florence would not have been possible.

Lasting Change

Variations alone, though, are not enough; they also have to be able to persist. One form of cultural persistence is through personal contact and tradition. Verrocchio was able to transmit his knowledge to the young Leonardo through conversation and personal demonstration. In this way Verrocchio could convey what he had learned over the years to his pupil. Similarly, the mathematician Isaac Barrow inspired his student, Isaac Newton, and the botanist John Henslow taught young Charles Darwin. Each pupil may in turn pass on what they have learned, allowing knowledge and traditions to be continually transmitted from generation to generation. Learning from others is not of course restricted to formal tutoring. It happens all the time as we interact and converse with each other.

Cultural persistence may also be mediated through artifacts. No matter how original Leonardo was, he would have had little impact if all of his paintings, drawings, and writings evaporated as soon as he finished them. Producing materially stable artifacts provides a very effective means for the longterm transmission of knowledge to others. Verrocchio and Leonardo could inspect and learn from the Pollaiuolos' *Tobias and the Angel*, because the paintings were durable. For the same reason, works from the Renaissance could continue to inspire artists in later centuries. Joshua Reynolds, who traveled to Italy to see such works, advised aspiring artists in 1769: "Consider with yourself how a Michael Angelo or a Raffaelle would have treated this subject: and work yourself into a belief that your picture is to be seen and criticized by them when completed." Similarly, Paul Cézanne gained inspiration from the old masters by regularly visiting the Louvre museum in the afternoon to observe and sketch their works. Looking further back, we are able to appreciate the cave art of Magdalenian societies completed more than ten thousand years ago (fig. 2, page 21). All of this is possible because of material persistence. The same applies to other artifacts like clothes, writing, or sculpture. With products persisting over time, people can learn from and build on the efforts of their predecessors.

Persistence may also be enhanced by making copies. Up until Leonardo's time, copying was still largely done by hand or with woodblocks. But with Johannes Gutenberg's invention of the printing press with moveable type in about 1430, copies could be produced far more effectively. By 1500, Gutenberg-style presses could be found in 236 towns in Europe, and had printed about twenty million books in various languages. Leonardo benefited from this printing bonanza; he acquired more than one hundred printed books on subjects ranging from mathematics, philosophy, and natural history, to poetry. Printing therefore played a critical role in allowing ideas to persist and spread. New copying methods have continued to be developed, including photography and electronic dissemination through the internet. As well as these forms of propagation, we may also generate copies by repetitive making. The mass production of clothes, cutlery, and cars all depend on machines that repeatedly carry out particular actions. Unlike printing, one car does not act as a direct template for another; cars arise by continually repeating the same process of manufacture.

All these forms of persistence—personal instruction, durable artifacts, and copying—allow cultural achievements to have lasting effects. Without persistence, every individual advance, no matter how wonderful, would be carried to the grave with its inventor and have little cultural significance.

Cultural persistence and variation are closely connected. Both revolve around the ability of humans to communicate, most notably through language. Communication is important for generating variation through exchanges among people, but also for the retention and maintenance of knowledge. Variation alone would create a frenzy of ephemeral ideas and actions, while persistence alone would lead to stagnation. It is through a balance between variation and persistence that effective cultural change becomes possible.

Cultural Reinforcement

Population variation and persistence provide key ingredients for cultural change, but the drive for change within populations as a whole comes about through a further ingredient: reinforcement. As their paintings were exhibited in Florence, the reputation of artists such as the Pollaiuolos, Verrocchio, or Leonardo could spread in the population. If every person who likes a painting tells two others about it then the message multiplies like an infection. And as their reputation spreads, artists can win more lucrative commissions which may further enhance their fame. Success breeds success. Even the achievements of Vincent van Gogh or Paul Cézanne, who were not seen as particularly successful in their lifetimes, were reinforced later by others who recognized and advocated their work.

Certain achievements are able to be reinforced in human populations because people can have common values; they often have similar likes and dislikes. As we saw in previous chapters, what we learn and do depends on what we value. We may value many things such as food, sex, comfort, goods, land, money, status, knowledge, and beauty. And because many of these values are shared in a population, achievements that meet those values tend to get reinforced and spread.

Not all individuals of a population share exactly the same values, however, due to variations in birth and upbringing. Some may value particular things, such as goods, money, or knowledge, more highly than others. Specialists may also value things that others are less aware of; when looking at the paintings of the Pollaiuolo brothers, Leonardo and Verrocchio no doubt saw and valued features that most others would not have appreciated. Human populations possess a diverse collection of values. Nevertheless, because many values are held in common among certain social groups, achievements that meet those values spread through reinforcement, at least within those groups.

What we value is also strongly dependent on cultural context. When Leonardo was thirty he moved to the city of Milan, ruled by the Sforza family. Unlike the Medici of Florence who were bankers, the Sforzas were military men who drew strength from their armed forces. When Leonardo wrote a letter introducing himself to the Duke of Milan, Ludovico Sforza, he pitched it with the Duke's military interests in mind:

> My most illustrious Lord, Having sufficiently seen and examined the inventions of all those who count themselves makers and masters of instruments of war, I have found that in design and operation their machines are in no way different from those in common use. I therefore make bold, without ill-will to any, to offer my skills to Your Excellency, and to acquaint Your Lordship with my secrets, and will be glad to demonstrate effectively all those things, at whatever time may be convenient to you.

Leonardo then went on to list the series of inventions he had devised that would put Sforza ahead in the military game. These included portable military bridges, methods of removing water from trenches when a siege was underway, methods for destroying fortresses, novel types of cannons, and designs for armored cars. The list reflected what Leonardo thought would be of particular value to Ludovico Sforza. And Sforza's interests in turn reflected the prevalence of military conflict among rival families and states at that time in Italy. Leonardo himself witnessed the military expulsion of the Sforzas by the French several years after he moved to Milan.

This is an example of how cultural values are not fixed; they depend on the context, social groupings, and prevalent issues of the time. Nevertheless, because they can be shared among many individuals in a population or social group, they lead to the reinforcement of ideas and achievements that meet those common values. Reinforcement is a key ingredient of cultural change; for it is through this process that particular achievements are driven to ascent in a population.

The Force of Competition

Success often leads to competition. As an idea or achievement spreads, the novelty eventually starts to wear off as it becomes commonly known. The achievement essentially starts to compete against itself because as the majority of a population grows familiar with it, the rate at which it spreads diminishes. Successful achievements also stimulate competition from others. When Verrocchio became aware of how well the Pollaiuolo brothers were doing, he took up painting so that he too might find a share of the lucrative market. The success of the Pollaiuolos even helped the competition. By carefully observing the Pollaiuolos' *Tobias and the Angel,* Verrocchio and Leonardo could benefit from their advances, and then produce an improved version. For the Pollaiuolos, having their achievements brought to the attention of others was a double-edged sword: the attention was a measure of their success, but it also provoked and assisted competition. Similarly, the success and fame that Leonardo was to later achieve stimulated the achievements of competitors, such as Michelangelo. This is another example of our familiar double feedback loop, with reinforcement enhancing competition (fig. 82). A valued achievement spreads through communication, but this success is self-limiting, because novelty wears off as knowledge is shared in the population, and competition is encouraged.

Cultural competition arises because of several limitations. There are limits to the size of human populations, to the number of achievements they value highly, and to the rewards that may be won. Florentine society only recognized a few artists as being truly outstanding, and rich families, such as the Medici, commissioned a limited number of works. Such limitations inevitably bring about competition. If the population was infinite, then the novelty of an achievement would never wear off because it could continually spread to new ears. And if commissions had been limitless and recognition inexhaustible, then there would have been little reason for Verrocchio and the Pollaiuolos to compete.

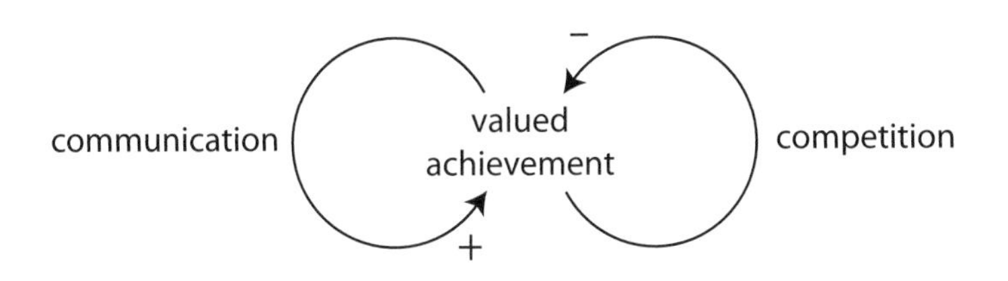

Figure 82. Interplay between reinforcement (positive loop) and competition (negative loop) for human culture.

Even when we might think that an individual is working purely because of curiosity or a search for personal satisfaction, competition tends to surface when success or recognition is at stake. Darwin spent many years developing his theory of evolution by natural selection without publishing his ideas. But when he received a letter from Alfred Russel Wallace presenting the same theory he grew alarmed, fearing that Wallace might get the credit. Darwin found an amicable solution and they published their findings together, but the story highlights how competition can lurk behind the scenes. Similarly, when Leonardo wrote about his secret knowledge of war machines in his letter to the Duke of Milan, he was clearly aware of competitors that could steal his ideas and so deny him and the Duke of their advantage.

Competition is sometimes viewed as a negative or undesirable aspect of humanity. Yet it plays a vital role in encouraging cultural change. Had Verrocchio not been competing with the Pollaiuolos, he might not have taken up painting, and Leonardo may not have developed as he did. Had Leonardo not aspired to come up with inventions or ideas that were better than those of others, perhaps he would not have been driven to achieve so much. Reinforcement and competition go hand in hand. Reinforcement allows valued achievements to spread, provoking the competition that in turn spurs on further achievements. Without some form of competition, culture would come to a standstill.

Cooperative Efforts

When Leonardo went to work in Verrocchio's workshop, it was beneficial to both parties. Leonardo learned important skills from his teacher while Verrocchio profited from his apprentice's help. The painting of *Tobias and the Angel*, with both artists contributing parts of the picture, embodies this cooperation. Being in the same place was an important aspect of this relationship. As with other cases of cooperation, physical proximity provides conditions that can encourage mutual assistance and sharing of benefits.

Cooperation plays a vital role in cultural change at many levels. Our ability to talk and communicate is a result of cooperation among humans. If you were brought up in complete isolation then you would not learn an elaborate language or have the ability to communicate effectively with others. And without the ability to communicate, you would be unable to benefit from others, and your achievements would be unlikely to persist and spread. Language and communication, which play such a vital role in cultural change, are only possible because humans talk and interact with each other in groups. Similarly, our propensity to exchange money and goods is a cooperative venture. Leonardo was sustained by commissions and payments from families like the Medici or Sforzas. Conversely, those who purchased his works benefited by owning prestigious paintings of high quality. Without such an economic setting of exchange and cooperation, Leonardo would not have been able to support himself and achieve what he did.

Further levels of cooperation also occur. As well as belonging to Verrocchio's workshop, Leonardo was a member of the artists' guild in Florence. This was one of many guilds at the time, that included guilds for cloth merchants, bankers, and physicians, each looking after the interests of its members. Leonardo was also a citizen of Florence, another grouping defined by common interests. These various groups developed through interplay between cooperation and competition. The cooperation between Leonardo and Verrocchio was promoted by competition—Verrocchio needed apprentices to produce works that would rival those of the Pollaiuolo brothers. And other groupings, such as guilds or cities, were also sustained through rivalry. Being primarily a painter, Leonardo was not ashamed of promoting his own medium compared to others: "Sculpture reveals what it is with little effort; painting seems a thing miraculous, making things intangible appear tangible, presenting in relief things which are flat, in distance things near at hand." Of course, those with expertise in other areas might dismiss painting just as Leonardo dismissed sculpture. The poet Lord Byron expressed his view on painting: "Depend upon it, of all the arts, it is the most artificial and unnatural, and that by which the nonsense of mankind is the most imposed upon." Such disparagement of competing groups is not restricted to the arts. The experimental physicist Ernest Rutherford, discoverer of the atomic nucleus, tended to minimize the contribution of his more theoretically inclined colleagues. He has been quoted as saying, theorists "play games with their symbols, but we in the Cavendish turn out the real facts of nature." As we have seen in previous chapters, cooperation can both engender and be sustained by competition.

The interplay between cooperation and competition was something that Leonardo clearly appreciated. In his writings on painting, he considered the various advantages of working alongside others:

> I say and am prepared to prove that it is much better to be in the company of others when you draw rather than alone, for many reasons. The first is that you will be ashamed of being seen in the ranks of the draughtsmen if you are outclassed by them, and this feeling of shame will cause you to make progress in study; secondly a rather commendable envy will stimulate you to join the number of those who are more praised than you are, for the praises of the others will serve you as a spur; yet another is that you will acquire something of the manner of anyone whose work is better than yours, while if you are better than the others you will profit by seeing how to avoid their errors, and the praises of others will tend to increase your powers.

For Leonardo, putting artists together resulted in a creative mix of competitive and cooperative interactions, and was clearly more effective than everyone working alone.

Cooperation and competition are partners in cultural change. Cooperation allows humans to learn and benefit from each other at a range of scales, from the ability of individuals to communicate to larger collective enterprises. And by promoting shared values and aspirations, it helps reinforce achievements. Competition plays a key role in sustaining cooperation while also providing a continual spur to further achievement.

A Cultural Mix

By bringing people together, cooperation also encourages ideas and abilities to be combined in new ways. We do not invent from scratch. We combine and build on what is already known. As Leonardo worked in

Verrocchio's workshop, he continually integrated his ideas and observations with what he learned from his master and the other assistants working around him. His achievements represented an original combination of skills and knowledge, rather than something that was entirely new. For Joshua Reynolds, "invention, strictly speaking, is little more than a new combination of those images which have been previously gathered and deposited in the memory." This is the principle of combinatorial richness, but now in the form of people, their ideas, and their achievements.

The role of combinatorial richness is most evident when we look at the media through which cultural achievements are propagated. We have only twenty-six letters in our written alphabet, but this does not stop us from expressing new and complicated ideas in poetry, literature, and science. When we arrive at a novel insight we do not feel the need to invent new letters, because the number of possible combinations with what we already have is almost limitless. The same applies to words. We may occasionally require some new ones, such as the term *gene*, coined by Wilhelm Johannsen in 1909. But the words we already have are usually sufficient: it is through combining them in particular ways that new ideas are expressed and communicated. Similar considerations apply to musical or mathematical notation. Painting is also a combinatorial activity. Artists can mix a few colors on their palette and distribute them in various combinations to wonderful effect. Combinatorial richness is evident at many levels in culture, from the arrangements of letters and words of a language, to juxtaposition of ideas or materials, or encounters among diverse people.

In previous chapters, the large number of possibilities arising through combinatorial richness was conveyed through high dimensional spaces. In evolution, populations journeyed through genetic space, in development an embryo took an odyssey through developmental space, and during learning a brain traveled through neural space. We can now think of the human species traveling though cultural space. Cultural space represents the vast range of cultural possibilities for humans. At any one time, we can think of the human species—all the peoples of the earth with their various interests and achievements—as representing one position in this unimaginably large cultural space. Over time, our species has journeyed through cultural space, taking a particular path as human habits and endeavors have become modified across the globe.

When viewing the journey of our species through cultural space, it is tempting to see it as a progression that led to our present culture. However, whether we view a change as progress or not depends on our particular values, and these may themselves change with time. Moreover, the journey of our species has not taken a simple linear path but has involved many complex twists and turns. It is therefore misleading to think of the human journey as being inevitably pulled toward the present. Even so, we can briefly summarize some of the key steps that took place along the way.

Even before our species *Homo sapiens* embarked on this cultural journey, related human species embarked on some preliminary voyages. The earliest human artifacts, stone chopping tools from Africa, date back to around two million years ago. They are thought to have been made by *Homo habilis*, a species with a smaller skull than ours, with about half our brain volume. The tools they left behind suggest that these humans were already able to think ahead. A stone tool is made with a future purpose in mind. It has value through the delayed reward it offers—the meat that it will help its user cut from its prey. This value would have been shared among individuals of the species, and assuming they could also communicate to some extent, tools and toolmaking could have spread in early human populations through reinforcement. The need to produce better tools may also have led to competition, and cooperation may have increased as one toolmaker learned from another. In the ensuing years humans grew better at making tools, producing a range of highly effective hand axes. This may have partly been because of evolutionary changes in brain capacity, but also because of manufacturing improvements transmitted through culture. Early human species were already journeying through cultural space.

Modern humans, the species *Homo sapiens*, are thought to have evolved about two hundred thousand years ago in Africa. From there they eventually started to spread like a weed across the globe, outcompeting relatives, such as *Homo neanderthalensis*. Why the dominance of *Homo sapiens* started asserting itself during this period is not clear. Developments in language and communication were no doubt involved, but exactly how *Homo sapiens* became the dominant human species is a subject of much discussion. Moreover, as cultural change involves interactions among multiple ingredients, it is probably a mistake to search for a single underlying cause. Whatever the reasons, by ten thousand years ago, *Homo sapiens* was the only human species remaining on the planet. Our species then lived on all continents except for Antarctica, forming a global population of a few million. It was around then that humans learned how to cultivate plants and animals. Seeds and livestock became valuable because of the promise of food they would yield in the future, and skills in agricultural production began to develop and spread. The development of agriculture represents a key step in the journey of our species through cultural space, because it provided stability and a surplus of food that enabled additional innovations to develop and spread.

With improved transport and farming techniques, humans began to form larger organized groups, or civilizations, about five thousand years ago along the fertile river valleys of the Nile in Egypt, the Tigris and Euphrates in Mesopotamia (modern Iraq), the Indus in Pakistan, and the Huang He (Yellow River) in China. And as civilizations cooperated and competed, with new ones also emerging, our journey through cultural space continued, exploring new directions with the help of innovations like writing and money. The ever increasing ability of humans to grow food and harness resources also meant that their numbers increased. By two thousand years ago the world population had reached about two hundred million. Population expansion has continued to the present day, spurred on in the nineteenth century by our greater ability to harness energy from fossil fuels, and in the twentieth century through the development of medicines, hygiene, and intensive farming. Our position in cultural space today reflects the habits, aspirations, and achievements of seven billion people. And of course our species is still on the cultural move.

The human species has taken a particular path through an unimaginably vast cultural space. What is it that continually drove us on?

Propelled by the Past

All of the principles we have covered so far in this chapter operate within a particular social context. The type of variation in human populations, its persistence, the values that reinforce it, the way people compete, cooperate, and combine their efforts all apply to particular social situations. One of these situations was the population of Florence in the fifteenth century. But these contexts, these social situations, are not fixed; they change through the very processes we have encountered. By responding to the Pollaiuolos' painting of *Tobias and the Angel*, Verrocchio and Leonardo modified the artistic context by creating their own version: their painting provided a new standard for others to react to. And the same story continues with other artists down the line, from Michelangelo to Monet. Even when someone reacts negatively to their precursors, such as Cubists rebelling against the rules of perspective, the previous works play a vital role in engendering that reaction. One context provides the seeds for the next.

Similar considerations apply to military achievements. Because cannons had already been invented and were in common use, Leonardo was able to devise particular improvements. As soon as improvements become known, all armies begin to use them, and then further improvements were in turn needed to gain a competitive edge. This is why Leonardo emphasized secrecy in his letter to the Duke of Milan. Each new military invention was not only based on previous ones, it also drove the process forward by changing the context. Similarly, as soon as a scientific theory becomes established, it provides the foundations for further investigations and scientific developments. This may lead to improvement of the theory or to its overthrow by a new and better theory. Contexts provide the seeds of their own change because they not only lead to particular responses, but those responses themselves change the context.

Cultural shifts are relational, always involving modifications based on what went before, with each step changing the context and standards, and so providing the impetus for the next. And this doesn't happen just in one place or discipline, it happens repeatedly across the globe in many areas of cultural activity, including the arts, technology, science, fashion, or commerce. It is our familiar principle of recurrence, applied to human endeavors.

These self-propelling shifts in cultural space not only involve the spread of new achievements, they also involve changes in some of the values that reinforce those achievements. Leonardo had several criteria for judging the success of a painter's work: "First you should consider the figures whether they have the relief which their position requires . . . . Secondly, you should consider whether the distribution or arrangement of the figures is devised in agreement with the conditions you desire to represent. Thirdly, whether the figures are actively engaged on their purpose." While appropriate to Renaissance art, with its emphasis on capturing reality and perspective, these values would hardly apply to Impressionist or Cubist painters that came later. What we value, or our criteria of success, may change as we respond to, and build on, what went before.

This does not mean that all values are continually changing. Painters through the ages no doubt shared some values, such as a love of painting and observation. We can discuss the history of painting as a coherent subject because of such basic commonalities. Each artist may react to their colleagues and predecessors, and introduce new elements to the field. Impressionism did not appear out of the blue, but emerged by building on what went before. Nevertheless, saying that a painting by an Impressionist like Monet is better than that of a Renaissance artist, such as Leonardo, would be considered a subjective statement. There is no fixed or common criterion that allows us to make such judgments, because the values through which we evaluate a painting may vary. We may even employ multiple values at once, and decide that both Monet and Leonardo are great artists, each in their own way. Fashions in clothing are another example of how some values may vary. The basic value of looking good may not change, while the particular types of dress we value may vary from one year to another. Some values are more slippery than others and alter more rapidly, driven by the process of cultural change itself.

Science is based on a relatively fixed set of core values that involve how well or simply theories account for observations. These values were not always prevalent. Leonardo continually emphasized the importance of observation in forming scientific opinions, rather than relying on hearsay and traditional beliefs. Indeed, it was during the Renaissance that our modern scientific values began to emerge. But once these values became established and firmly embedded in culture, they allowed scientific achievements to be judged through a relatively fixed set of criteria. We can discuss progress in science because we can ask which theory provides a better or simpler explanation of what we observe. This leads to improved matches between theory and observation. Similarly, we can discuss improvements in technologies like communication, because we have accepted measures we can refer to, such as the speed at which information travels. In having broadly fixed criteria against which things are judged, science and technology are similar to processes like evolution and predictive learning. In evolution, organisms or genes are "judged" according to the criterion of reproductive success, while in predictive learning, neural changes are "judged" according to the criterion of reward delivery.

A further feature of recurrent cultural shifts is that they may feed back to influence the effectiveness of the principles they are based upon. One of the main trends in our cultural journey has been an increase in the speed and range of communication and transport. Our ability to exploit horses and camels played a key role in extending the range and interactions of early civilizations on land. Similarly, innovations in shipping and navigation played a key part in allowing transport of goods and ideas among diverse countries. These developments were driven by the value humans placed on land and goods. But by linking diverse populations more effectively, these developments also resulted in the improved effectiveness of our ingredients of cultural change. They increased variation by increasing the size of the population that can interact; they allowed achievements to spread further afield through persistence and reinforcement; they increased the range of competition and cooperation; and by bringing diverse people in contact with one another they promoted further levels combinatorial richness. The collective effect was to accelerate further technical developments in transport and communication, eventually leading to trains, cars, planes, telephones, and the internet. Each of these innovations was built on what went before—the early name for a car was a horseless carriage, and a radio was known as a wireless. These are strange names unless you know their history. And the net effect of these improvements in communication and transport was that the world shrank while the human population expanded, accelerating the process of cultural change by feeding back to improve the effectiveness of the principles it is based upon. The ever-increasing pace of cultural change we witness today is the outcome of such feedback.

A Cultural Recipe

We have seen how the process of cultural change operates according to the same recipe encountered throughout this book. The driving force for change is the double feedback loop between reinforcement and competition, with success both promoting itself and bringing about its own limitations. These loops are fueled by a balance of population variation, which continually generates new ideas and juxtapositions, and persistence, which allows achievements to be maintained and spread through the population. Cooperation also plays an essential role by allowing people to benefit from each other's skills. This both promotes achievements within groups and leads to further levels of competition among them. By bringing people and ideas together, cooperation also leads to an enormous number of combinatorial possibilities, creating a vast cultural space through which our species can move. This movement is self-propelled, continually redefining its own context through the principle of recurrence.

These are the same principles and interactions that we found at the foundations of evolution, development, and learning. In evolution, they involve populations of individuals reproducing in an environment. For development, the principles concern populations of molecules or cells within a growing embryo. For learning, they involve populations of neural firings in the brain. With cultural change, our recipe applies to populations of interacting human beings. For each system, we see populations propelling themselves through vast combinatorial spaces: genetic, developmental, neural, and cultural. Cultural change is our fourth instance of life's creative recipe.

We should be careful though not to try to push these resemblances too far. Cultural change is distinct in many ways from the other processes. A creative act is very different from a mutation or molecular collision, and the patterns of cultural transmission involving language, artifacts, and copying are not the same as the laws of genetic inheritance or synaptic modification. There have been attempts to link culture and related biological processes more precisely. For example, a unit of transmission called the meme has been proposed as playing a similar role in culture as a gene plays in evolution. A meme is an idea or activity that replicates and is transmitted from one human to the next, just as a gene may be passed from parent to child. While the notion of memes may be useful in emphasizing certain parallels between evolution and cultural change, it also invites confusion. What, for example, is the precise equivalent of natural selection that leads one meme to increase at the expense of another? The answer is far from clear, leading the geneticist Jerry Coyne to conclude that the problem with memetics is that it "seems completely tautological, unable to explain why a meme spreads except by asserting, *post facto*, that it had qualities enabling it to spread."

It is important to stand back to view the relationship between culture and the other processes at an appropriate level of abstraction. Through the common overall principles of life's creative recipe, we can see their similarity in form. But being aware of formal similarities does not mean we should seek one-to-one correspondence between components. We may recognize that chess and war are both territorial enterprises without trying to match squares on a chess board with units of land in a battle. Indeed, trying to find a corresponding unit of land would be a misguided effort that distracts rather than helps with understanding these processes.

I have tried to show how life's creative recipe provides a useful and appropriate level of abstraction, emphasizing similarities in form that apply to different types of living transformation. We are then able to see that there is a common set of principles and interactions at the heart of each process. We can also appreciate that as humans we have had the privilege of being part of four remarkable journeys: the journey through genetic space that has led to the human species; the journey through developmental space that led to our birth and growth; the journey through neural space that has led to our thoughts and actions; and the journey through cultural space that allows us to enjoy the ideas and achievements of others. We are both passengers and participants in these four great voyages of life.

As well as having a similar form, these journeys are also connected in other ways. It was through the earlier processes of evolution, development, and learning that the basic ingredients of cultural change arose. Our cultural journey follows from the other three. But there is also a relationship that goes in the other direction. Our scientific understanding of evolution, development, and learning is itself a cultural product. It is through culture that we view all living transformations. To gain a better understanding of this dual relationship between culture and the other transformations, we need to look more closely at how our four journeys are connected.

TWELVE

The Grand Cycle

HUMANS ARE GREAT STORYTELLERS. It is through stories that we often communicate our knowledge and understanding. We like our stories to have distinct structures, with clear beginnings and ends, and we usually like things to happen for particular reasons that we can lay out in sequence. These considerations also apply when it comes to the story of ourselves. In tracing the story of humanity we naturally ask when it began, and what makes humans unique. We often seek answers in terms of defining qualities like language, imitation, the ability to plan ahead, the human soul, or creativity. Yet throughout this book, when we have looked into the origins of living transformations we have not found clear starting points or defining essences. There are no obvious beginnings or single qualities that account for evolution, development, or learning. Rather, these transformations arise through a constellation of ingredients coming together and interacting with each other. Instead of clear starting points we have found loops and interactions.

The same applies to cultural change. Our outlook today cannot be traced to a single cause but depends on a recipe with many interacting ingredients. Instead of a particular starting date or defining essence, the origins of human culture are to be found in how its various components arose and came together. Let's look at where each of the principles underlying cultural transformation came from.

Cultural Origins

The variation of humans results from differences in birth and upbringing. The brain we are born with is influenced by the particular genes we inherit and how they modify our development in the womb. After birth, further individual characteristics are acquired as we encounter various situations and learn from them. Through these encounters, our brain continually builds upon expectations and discrepancies, leading to our becoming individuals with particular ways of seeing and doing things. The variation that allows cultural change in human populations can be traced to the character of brains and their interactions, which in turn reflects the interplay between evolution, development, and learning.

Similar considerations apply to cultural persistence. Transmission of knowledge and expertise from one person to another depends on our abilities to communicate. This skill in turn stems from how each of us develops and learns. Our brain enables us to learn languages, gestures, and actions, and thus communicate effectively with others. We may also learn to make tools or other artifacts that persist, and help to transmit knowledge from one person to the next. Cultural persistence is rooted in the way our brain evolved, develops and learns, allowing us to communicate successfully and with lasting consequences.

Our ability to communicate also plays a key role in cultural reinforcement. Through communication particular ideas and achievements can spread in a population. But what is disseminated also depends on what we value. We are born with particular values embedded in the neural connections of our brains. These values are rooted in our evolutionary past. Our basic wishes for comfort, food, and a partner arose because they tend to promote survival and reproduction. Through learning, these and other inborn values are built upon in various ways. We may come to value money, fame, knowledge, or a beautiful painting, but whatever the values we acquire, they are rooted in our biological inheritance and then modified through learning. As with previous principles, cultural reinforcement is grounded in evolution, development, and learning.

The principles of cultural competition and cooperation have similar biological roots. Having evolved as social animals, humans are endowed with a fine balance of competitive and cooperative drives. Our competitive nature stems from the need to promote our own survival and reproduction.

By competing for food, comfort, or attention we may increase the likelihood that we survive and leave offspring. Cooperation within a social group is also important for we may benefit through mutual assistance. By cooperating with our partner we may increase the chance that our children survive and prosper. Competition and cooperation interact in all sorts of ways during the process of culture change, but they are always rooted in our biological past.

The same applies to the principle of combinatorial richness. Brains are structured to operate in a combinatorial way, integrating information by combining multiple inputs. For humans, these inputs may include information from other people, their sayings and doings. As our neural networks feed off each other, these social inputs lead to some neural connections growing stronger while others weaken. Through this process our brain becomes modified, allowing us to benefit from the achievements of others and arrive at new combinations in the form of language, artifacts, or ideas. The vast cultural space in which the human species travels is rooted in the combinatorial way our brains develop, function, and interact.

Finally, the principle of recurrence is also grounded in our biological makeup. From an evolutionary point of view, it doesn't pay to be satisfied. It is better to continually search for actions that might increase the chance of survival and reproduction. As in many other animals, our brains are structured to seek the best course of action among the options available. Even if conditions improve, complacency never pays; opportunities might be missed and we may risk others doing better than ourselves. For a given set of options, we therefore tend to seek the one that best matches our values: the best food, the best partner, the best artifact, the best means of communicating, the best way of getting around, or the best explanation. Both the available options and what we value may change with time, but our search for the best does not. Even a reclusive monk may seek the best place to meditate or the best way of life. A lack of interest in looking for the best, not caring about what happens, is regarded as an illness, a sign of disengagement or depression.

As we saw with TD-learning (chapter 7), we learn in a relational manner, always adjusting our expectations by learning from discrepancies. Because of this, as certain options or values are reinforced and become established, new ones may appear and lead to additional rounds of reinforcement and competition. This continual shifting of expectations and values is what keeps culture on the move. If we reached a point where everyone was completely satisfied then perhaps culture would come to a halt. But this is very unlikely because we have a natural tendency to grow dissatisfied with some aspects of life: our brains continually search for discrepancies, and actions that help to resolve them. Moreover, if a stagnant pocket of human culture were to develop, it would be vulnerable to being overrun by a competing group of less satisfied people that seek new territories or resources. The principle of recurrence, the ever shifting context in cultural space, is grounded in our biology, our continual drive for the best course of action among the options available.

All of the ingredients for cultural change are grounded in our biological past. We would come across many of them in other social animals, from termites to dogs. But with the evolution of *Homo sapiens*, the ingredients came together and interacted with particular force. Our abilities to learn, communicate, and interact are much greater than those of any other creature. A termite mound is a magnificent collective effort, but its basic form may only change on an evolutionary timescale through alterations in the instinctive reactions of its builders. By contrast, our houses and cities are very different from the dwellings of our ancestors living ten thousand years ago. This is not due to a change in our genes, but because of the cultural transformations that arise from our abilities to learn, communicate, and interact. As our cultural journey has progressed, our advances have also fed back to enhance the various ingredients it is based upon. The development of more effective means of communication and travel, for example, enhanced population variation, persistence, reinforcement, competition, cooperation, and combinatorial richness, further catalyzing our recurrent movement through cultural space.

We have seen how evolution, development, and learning are three instances of life's creative recipe that frame each other. All three processes have a similar form but they are also connected through history, with development embedded in evolution and learning embedded in both other processes. Cultural change can now be seen as a fourth instance of the recipe framed by its predecessors. It not only resembles the other transformations through a common creative recipe, it is embedded within them.

But there is a further twist to this tale. Our understanding of evolution, development, and learning is itself a product of culture. The theories of Darwin or Turing are outcomes of our cultural heritage over the centuries. Our fourth instance of the recipe is special in one particular sense. For human culture is not only framed by the other three processes; it is through culture that we have come to view all the others. It both frames and is framed by them all. It would be as if Sinbad when telling his story to his guests started to read aloud the book *One Thousand and One Nights* that contains his own story.

Culture as a framing system applies not just to biology, but to the whole of science. The theories of Newton and Einstein are just as much a product of culture as those of Darwin or Turing. How can we reconcile this double aspect of culture? What should we consider as primary, culture as the framer of science or science as the framer of culture? To answer this question, I want to look at where some of our most basic ideas about the world come from.

Possible Worlds

We commonly describe space as having three dimensions, containing material objects with length, breadth, and depth. We can imagine objects with fewer dimensions, but they would not have a material existence; a two-dimensional object would be infinitely thin and therefore insubstantial. But it is not so easy to think about objects having more than three dimensions. We can't readily imagine four-dimensional objects. Is this limitation a feature of our outlook or the physical world around us?

Let's first look at the various means by which we evaluate objects. As you hold an apple in your hand you may appreciate many of its aspects, including its color, smell, and shape. If the lights go out, the apple disappears from view, yet its scent and shape is still evident to you through your senses of smell and touch. You would say the apple is still there but that you just can't see it. Similarly, if you could no longer smell the apple for some reason, you would not judge that the apple had gone away because you would still be able to see and feel it. The situation is very different, however, if you imagine the apple no longer resisting your grasp. If you saw your fingers pass through the apple as you tried to close them around it, then you might conclude that the apple had become a phantom that no longer really existed. The way an apple feels and meets our contact seems to be integral to it, while attributes such as visual appearance and smell are more incidental.

From a biological perspective, the greater significance we attribute to some of an object's aspects over others is related to what matters most for our survival and reproduction. The most important thing for us is the pattern of contacts we make with objects in our surroundings. We eat by making contact with food, we reproduce by making contact with fellow humans, and we prolong our lives by avoiding contact with things that might kill us. Our sense of touch provides us with our most immediate access to contact. When we touch something, we feel it is real because the contact may do us real harm or good. Senses like vision are more indirect. We cannot eat or reproduce by seeing. Vision is only important insofar as it helps us promote or avoid particular contacts. Seeing may help us make contact with food or a mate, and avoid contact with a predator. We tend to think of touch as the primary arbiter of reality, a judge of what really counts, while seeing has a secondary significance. If we lived like plants, this might be otherwise. For then, being exposed to light would have a very direct consequence on our survival, because we would depend on it directly for food through photosynthesis, and visual qualities might seem more fundamental.

Some aspects of vision, such as the shape of an object, are strongly correlated with touch: we may both see and feel that an apple is round. These shared aspects contribute to our three-dimensional view of objects. Vision and touch collaborate in telling us that three dimensions are all we need to describe the extent of an object. Visual aspects that do not correlate with touch are not part of this description. The color of an apple is not directly relevant to how it feels, so our three dimensions are colorless. (They are scentless and noiseless for similar reasons.)

A good way of conveying four dimensions is to incorporate an additional quality into the way we experience contacts. I am going to use color as an example. Let me introduce rainbow world. Rainbow world is like ours except that each object has a single color (strictly speaking, a color interval), lying somewhere along the visible spectrum. Some objects are red, others yellow, and so on. The strangest feature of rainbow world is that objects only collide with each other if they have the same color. So, a red person can bump into a red table but can walk through an orange one. The red person feels only red things through contact, even though he or she can see objects with other colors. This applies whether the lights are on or off. A red person would still bump into a red table in the dark, they just would not be able to see that it was red. A further feature of rainbow world is that humans and other animals may change their color. If a red person wants to avoid bumping into a red table they may either decide to walk around it or change their color to orange and walk through it.

If we had evolved in rainbow world, we might naturally think of objects as having four dimensions. They would have the three dimensions of spatial extension, and the fourth dimension of color. We would become familiar with moving around in both extension space and color space to find our food and mates. To eat an apple, we would not only need to move toward it, but also change our color to match it. And if we wanted to avoid a predator, we would have the options of either changing color or running away. Of course the predator could in turn change its color to match ours, following us in color space just as it can follow us in extension space. Brains would become expert at four-dimensional thinking.

You might think that color here is acting rather like time, as we sometimes treat time as an extra dimension. But time is different because we cannot alter it by our own efforts and feel the consequences. We can of course go back in time by remembering. But when we do, we do not collide with the objects we remember. If you knew that someone had once been seated where you are now sitting, then remembering that event does not hurl you forcibly against them. By contrast, rainbow world is genuinely four dimensional because we can change color and feel the consequences, just as we can change our spatial position to make or avoid contact.

Examples like rainbow world suggest that our inability to think in four dimensions is not due to an intrinsic limitation of brains. Instead, this inability reflects the way we interact with the world around us. If our world operated according to the rainbow world rules, then we would have evolved, developed, and learned to think in four rather than three dimensions. Of course we don't live in rainbow world because objects with different colors don't have a habit of passing through each other. We live in a world in which our pattern of contacts can be accounted for with three dimensions. The origin of our outlook is not to be found purely in our brains or in the world around us, but in the way the two interact.

We may arrive at the same conclusion in a different way. Newton's theory of gravity has some limitations. As Einstein realized, the theory starts to break down when bodies approach the speed of light, and it fails to account correctly for the effect of gravity on light itself. Einstein addressed these problems with his theories of special and general relativity, which challenge some of our most cherished assumptions about space and time. We normally think of space and time as being independent of each other, with time passing in the same way regardless of how fast we move through space. Yet according to special relativity, time and space are intimately connected so that time may pass differently for objects that are moving relative to each other. This only becomes perceptible when objects approach the speed of light, so at our normal scale of existence such relativistic effects are negligible. Similarly, the effects of gravity on light, as described by general relativity, become most evident when masses are very large, as with a black hole where gravity is so strong that light cannot escape.

Relativity with its notion of space-time is more abstract and harder for us to grasp than Newton's theory, but this may be due to how we interact with the world rather than any inherent abstractness. Our brain is organized to cope with the objects around us, which typically do not approach the speed of light in their movement relative to us (light travels at about one billion kilometers per hour) and therefore these behave to a very good approximation as if space and time are independent. But this would not apply if we commonly operated at a different scale. As the astronomer Martin Rees writes: "An intelligence that could roam rapidly through the universe—constrained by the basic physical laws but not by current technology—would extend its intuitions about space and time to incorporate the distinctive and bizarre-seeming consequences of relativity." To such an intelligence living at a grander scale than ours, the theory of relativity might actually seem more straightforward than our notion of space and time as independent.

Similar stretches of the imagination apply to the world of the very small. Matter is made of countless atoms. It is natural to think of atoms as being like everyday objects except that they are extremely small, like miniscule ping-pong balls. This view, however, was overturned by the discoveries of quantum mechanics. At the scale of atoms and their component particles, such as electrons, assumptions about everyday objects no longer apply. You cannot, for example, know exactly where an electron is while also knowing how fast it is moving. It is simply impossible to stop an electron in its tracks and say "here it is." The more certain we become about where the electron is, the more uncertain we become about how fast it is moving. We start to talk in terms of probabilities rather than certainties, and matter takes on a fuzzy or jittery aspect. The fuzziness diminishes as the mass of an object increases, so by the time we get to visible objects like apples, which comprise massive constellations of atoms, the degree of uncertainty becomes negligible. We can point to an apple lying at a definite place on a table, because the level of uncertainty for something with the mass of an apple is so tiny that it is imperceptible. The quantum effects still exist in the apple but they remain insignificant unless we zoom into the scale of its constituent atoms. Just as our concepts of space and time can break down due to relativistic effects when speeds or masses grow very large, our familiar notions about objects and matter collapse due to quantum effects when we move down to the atomic scale.

This does not mean that there are three fundamental types of physical reality, one each for the small, medium, and large scales. Rather, it reflects way in which we normally interact with the world. As humans, we acquire concepts that allow us to deal with objects that we can readily see and touch, like apples or mountains. The notion of objects with defined locations in space and time is an extremely effective framework for coping with the world at this scale. Our everyday notions start to become less appropriate, however, when we move to scales that lie far away from our norm, whether very large or very small. To deal with these scales we need to stretch our familiar concepts, often in a way that can be uncomfortable or disconcerting. Because our ideas are based on interacting with the world at a particular scale, we are forced to think in a way that seems more abstract as we try to encompass broader scales.

Examples like rainbow world, relativity, and quantum mechanics highlight what many philosophers have also pointed out: we cannot access our world directly, but are always forced to look at it through particular frameworks. These frameworks are not arbitrary, but reflect the both the world and our interaction with it. Our notion of three-dimensional space has to do with the way we interact with the particular world we find ourselves in. If the world was different, we would also look at it differently. The theories of relativity and quantum mechanics are also frameworks for looking at the world. They reveal how the world is not as straightforward as our everyday interactions with it might suggest. Nevertheless, these frameworks are again arrived at by us interacting with the world, for it is through experiment and observation that scientists devised these theories. We cannot cleanly separate our view of the physical world from the ways in which we interact with it.

The same applies to our scientific explanations of evolution, development, and learning. We cannot describe these processes other than through the cultural frameworks that are based upon them. This does not mean that our scientific viewpoints are arbitrary. Rather, they reflect both the structure of the world and our interactions with it. The relationship between our four instances of life's creative recipe is not a simple one way chain from science to culture, or from culture to our science; it involves a two-way interplay between our views and the processes that gave rise to them.

Nature's Self-Portrait

Cézanne was a man obsessed. Some of his obsessions were rather unfortunate. He developed a terrible fear of being touched, as Émile Bernard came to realize after he made the mistake of trying to help him from falling one day. But his obsession with apples is something for which we are eternally grateful. Cézanne painted more than thirty still lifes with apples (see fig. 83, plate 14, for example). According to the critic Gustave Geffroy, Cézanne proclaimed: "I will astonish Paris with an apple." His aim was not simply to copy apples but to use these humble subjects to explore relationships in color, tonality, and form. Cézanne is reported to have said: "To paint from nature is to set free the essence of the model. Painting does not mean slavishly copying an object. The artist must perceive and capture harmony from among many relationships." Each artist may capture these harmonies and relationships in their own way. Compare Cézanne's painting with Renoir's treatment of a similar subject (fig. 84, plate 15), done at around the same time. Renoir's apples seem softer than Cézanne's and lack their strong sculptural quality, because he saw and painted them in his own way.

Scientists also portray apples in several ways. The apples of physics come in several varieties. There is Newton's apple, which falls according to the law of gravity, and Einstein's relativistic version embedded in space-time.

Figure 83. *Still Life with Quince, Apples, and Pears*, Paul Cézanne, c. 1885–87. See plate 14.

There is also the apple of quantum mechanics, comprising numerous particles that behave in a bizarre and fuzzy manner when we look closely. Then there are the various apples that arise through life's creative recipe: the apple that has been honed over many generations of natural selection, the apple tree that develops from a small pip, and the apple that we learn to appreciate through our neural frameworks.

All of these apples are connected with each other. The evolutionary apple is grounded in physics: organisms and their environment are made of matter, subject to its physical laws. The developing apple is rooted in evolution, for it is through the story of differential reproductive success

Figure 84. *Apples and Pears*, Pierre-Auguste Renoir, c. 1885–87. See plate 15.

that the development of pips into apple trees arose. The apple we see is also based on all of these predecessors as our interpretation stems from the way our brain evolved, develops, and learns. Finally, we have Cézanne's apple, a product of culture embedded in all the others.

But connections also run in the opposite direction, from the cultural apple back to the others. Our culture, including our scientific outlook, provides the frameworks through which we view the world. This means that we cannot ground our world view purely in physics, because as soon as we describe the physical world we are doing so through a particular cultural framework. Nor can we ground our world view purely in culture because our outlooks do not float alone but are embedded in the way we evolved, develop, and learn. Instead of a linear story with a simple beginning, we have a two-way interaction in which frameworks and what they frame are dependent on each other. Not being able to identify an absolute starting point may seem unsatisfactory. But perhaps this is because of yet another feature of our cultural heritage. We like stories to have a beginning, middle, and end. But this is not the way that the world, and our place within it, is structured.

The interdependence between frameworks and what they frame did not begin with humans. Every microbe can be thought of as a framework that captures relationships in its surroundings. The single-celled alga *Chlamydomonas* has Rubisco proteins that fit the shape of carbon dioxide molecules, allowing it to fix carbon. It also has whiplike flagella and a responsiveness that allow it to swim toward light. These adaptations allow the microbe to match particular aspects of its world. This matching arises through a continual two-way interaction between organism and environment. Organisms frame their environment in particular ways, and the environment then feeds back to influence which frameworks are favored through natural selection. The journey of populations through genetic space is intimately linked with the environment around them. Through this process, each species has evolved with its own constellations of features, representing a distinct way of capturing relationships in their surroundings, just as each artist has their own perspective on the world. This matching between organism and environment, between framework and framed, arose through evolution, our first instance of life's creative recipe.

Further frameworks arose when evolution spawned development, the second version of life's creative recipe. As multicellular organisms evolved, they could capture aspects of their surroundings at a grander scale. The bladder wrack captures sea motion, gravity, and the way light falls, by growing into a particular shape with various cell types—it carves up the world by carving up itself. As well as capturing spatial relationships, organisms also capture patterns in time. A weed may flower according to day length, the leaf of a Venus fly trap closes if an insect walks across it, and a sea slug may come to ignore being repeatedly touched. The journeys of embryos through developmental space, spawned by evolution, are intimately connected with their surroundings. Each multicellular plant and animal provides a particular portrait of the world that has arisen through evolution and development.

Responses to the environment have become particularly elaborate in animals which continually alter what they experience through their own

Figure 85. *Print Gallery*, M. C. Escher, 1956.

movements. Many of these creatures develop with nervous systems and brains that allow them to learn through the sequence of events they encounter. They are able to predict what is likely to be rewarding or punishing, and then modify their actions accordingly. They learn to calibrate their actions against their effects and negotiate the world around them more effectively. By exploring and probing the world, they anchor themselves within it, framing themselves and their surroundings at the same time. And by continually resolving and building upon discrepancies, they may arrive at new ways of seeing the world, finding new frameworks that capture what is around them. Learning represents another way of carving up the world through neural frameworks being modified according to experience.

The interdependence between the framework and the framed is not unique to culture; it is found in all instances of life's creative recipe. Evolution, development, and learning lead to arrangements of matter, known as organisms, which capture relationships about matter in their surroundings (including other organisms). Organisms are matter that frames itself. Life's creative recipe provides the general principles through which such self-framing arises. It is a recipe for self-depiction, a recipe through which the world portrays itself in various ways. With human societies, self-portrayal has been taken to a new level, to the way organisms conceive of their own origins and place in nature. This self-portrait reflects some of our own peculiarities as human beings. But the picture it portrays is not arbitrary, any more than a self-portrait by Rembrandt is an arbitrary collection of brush marks. Instead, the situation is like the lithograph by M. C. Escher showing a man looking at a picture of which he is a part (fig. 85). Like Escher's man we can never step out of our picture, but this does not mean that we cannot contemplate and try to understand the fascinating world of which we are an inseparable part.

Acknowledgments

A tree trunk does not start bare, but becomes so by shedding many branches. This might seem wasteful—why produce branches only to lose them? But these branches serve a critical function in the sapling by allowing it to harvest energy and grow. It is only when the tree achieves a certain stature that the lower branches start to be overshadowed and become dispensable. In the same way, many words and pages have had to be shed to produce this book. Little remains of the first drafts yet they played a vital role in allowing ideas to be tried out, explored, and built upon. I am grateful to many people for helping me with this process—pruning here or encouraging growth there, and so helping to formulate the ideas that eventually grew to form this book.

In particular, I would like to thank my colleague Andrew Bangham for his encouragement and stimulating discussions from the earliest stages of the book. His programming skills together with those of Andy Hanna were also critical for developing ideas on recognition and the software for portrait analysis used in chapter 9. I am also deeply indebted to Peter Dyan for his generous help and advice with the chapters on learning, and to my parents, Doris and Ernesto Coen, for their tireless support and critical reading of the many drafts I sent them.

I would also like to thank Graeme Mitchison for his very helpful comments and encouragement, Roger Carpenter for critical guidance on the science of eye movements, Chris Frith for helpful comments on cognition, Alicia Hidalgo for her general help and advice on developmental neurobiology, Jonathan Miller for stimulating conversations on recognition and cultural change, and Przemyslaw Prusinkiewicz for many helpful and enjoyable discussions. I am grateful to Richard Kennaway for computing the deformed Cézanne and to Ellen Poliakoff for the data on eye tracking. Thanks also for helpful input from Katie Abley, Dennis Bray, Antonio Cuadrado-Fernandez, Veronica Grieneisen, Martin Howard, Stan Marée, Marie Mirouze, and David Stern. I would also like to thank my agent Peter Tallack for helping to shape the argument in the book, my editor at PUP, Alison Kallet, for her wisdom and providing gentle steering when necessary, and copyeditor Sheila Dean and production editor Beth Clevenger for their helpful comments. Of course, any errors or misinterpretations made are entirely my own.

Thanks also to the members of my lab and personal assistant, Georgina van-Aswegen, for their support throughout the many periods book writing. And finally, I could not have written this book without the continuing support and understanding of my wife Lucinda and children Pip, Timmy, and Susie.

Notes

Given the broad range of this book, it is only possible to provide a selective set of notes and references. I have indicated those references that would be accessible to the general reader as distinct from more specialized technical references.

INTRODUCTION

Page

2 *This idea, however, was later shown to be misguided.* A thorough and reasonably accessible description of Haeckel's ideas has been given by Stephen Jay Gould (Gould 1977).

2 *A more recent attempt at bringing together evolution and learning was Gerald Edelman's theory of "neural Darwinism."* For an accessible account of neural Darwinism see Edelman 1994. A technical critique of Edelman's idea has been given by Francis Crick (Crick 1989).

3 *The origins of chess can be traced back to the game of chatrang played in Persia in the fifth to sixth centuries, which in turn may have derived from the Indian game of chaturanga.* The classic book on the history of chess is Murray 1913, but for a more recent and accessible account see Shenk 2007.

4 *It is thought that life on Earth originated about 3.8 billion years ago, and by 3.5 billion years ago our planet was populated by a diverse collection of single-celled creatures.* For accessible accounts see Knoll 2003 and Lane 2010.

8 *The sixth-century Chinese art critic Xie He came up with six ingredients that he thought were important for defining the quality of a painting.* For more information on Xie He and his relation to Chinese painting see Acker 1954.

10 *The Art of Genes* (Coen 1999).

10 *I came across an insightful essay written in 1990 by computer scientist Christoph von der Malsburg, in which he identified three basic principles that were common to self-organizing systems.* These principles are: (1) fluctuations self-amplify, (2) limitation of resources lead to competition between fluctuations, and (3) fluctuations cooperate (Von der Malsburg 1990).

CHAPTER ONE

Page

13 *All cultivated apple trees are thought to have descended from natural populations of* Malus pumila *in the Tian Shan region of Central Asia.* For an accessible account of apple history see Juniper and Mabberley 2006.

14 *Traditional accounts of evolution by natural selection invoke three main principles.* For some examples see Howard 1982, and Gould 1996. An accessible, overall account of evolution has been given by Ernst Mayr (Mayr 2002); for a textbook on the subject see Barton, Briggs, et al. 2007.

15 *This exceptional behavior of water is informative, however, because it tells us that expansion upon heating is not fundamental to changes in the states of matter.* For a comprehensive discussion of water and its peculiar properties see the readable book *H2O,* by Philip Ball (Ball 2000).

25 Quotation from Darwin 1958, pp. 42–43.

CHAPTER TWO

Page

34 Quotation about Cézanne from Vollard in Smith 1996, p. 56.

36 *But when faced with severe competition, it is often advantageous to combine forces and work together, as long as this is of mutual benefit.* A good but slightly technical description of key transitions of evolution involving cooperation at various levels is given in Smith and Szathmáry 1995.

40 Quotation from Sacks (Sacks 1995), p. 108.

43 *This is an astronomically large number, much larger than the number of organisms that have ever inhabited earth.* For other readable discussions on the large number of genetic possibilities see accounts by Richard Dawkins, Daniel Dennett, and Denis Noble (Dawkins 1986, Dennett 1996, Noble 2006).

51 *Consequently, our population clouds rarely stay still, but instead keep moving as new contexts create further challenges.* There are cases where some aspects of a species, such as the shape and appearance of its individuals, may not change over extended periods of evolutionary time. Even in such cases of stasis, however, there may be numerous changes in underlying processes that are not visible to us, such as resistance to disease.

53 *Yet all of these languages share common roots—they all trace back to a common ancestral tongue, thought to be spoken by the early founding tribes of Europe.*

An accessible account of language and its origins has been given by Steven Pinker (Pinker 1994).

56 *The higher the dimensionality of our system, the more likely it is that such trade-offs arise because there are more possible options to explore.* The importance of high dimensionality in promoting species divergence has been emphasized by Sergey Gavrilets (Gavrilets 2004).

58 *The ecosystems of our world reflect this balance between the production and extinction of species.* A discussion of how competition at the species level could be related to ecosystem diversity has been given by Stephen Hubbell (Hubbell 2001).

58 *As each population cloud moves, it influences other clouds, altering their context and nudging them in new directions.* This continual shifting of goal posts at the ecological level is sometimes called the Red Queen Effect and is nicely discussed by Matt Ridley (Ridley 1993).

59 *It is unlikely, however, that there was a clearly defined beginning because, as we have seen, evolution involves numerous feedback loops and the nature of loops is that they do not have clear starts.* Accessible accounts of the origin of life have been given by Andrew Knoll (Knoll 2003) and Nick Lane (Lane 2010).

CHAPTER THREE

Page

62 *Throughout his life, Turing was interested in discovering simple mathematical relationships that lie behind patterns.* For a good and readable biography of Alan Turing see Hodges 1992.

62 *These rules were first formulated by Alan Turing in 1952 to show how molecules could organize themselves.* Turing's classic paper on developmental patterning is Turing 1952. A more recent review of the modern relevance of his ideas is in Kondo and Miura 2010.

63 *If you place a tea bag in a cup of cold water, the water starts to turn brown as the tea dissolves and spreads from the tea bag to the rest of the cup.* The example of diffusion in a tea cup is also given by John Tyler Bonner in a small, readable book about the importance of size (Bonner 2006).

66 *Another way of achieving the same type of result involves a slightly different form of competition.* The notion of patterning through activators and inhibitors is nicely described by Hans Meinhardt in his book on shell patterns (Meinhardt 1998).

68 *In the past few decades, however, we have gained a much better understanding of how development actually takes place.* Some accessible accounts of development are given in Coen 1999, Carroll 2005, and Nüsslein-Volhard 2006. A useful textbook on development is Wolpert and Tickle 2011.

68 *The positioning of the partition depends on an interaction between two proteins, called MinD and MinE.* A technical review of the Min system for cell division is in Kruse, Howard, and Margolin 2007.

73 *Let's look at the formation of a particular type of cell, called a neuroblast, in a developing fruit fly.* I have given a simplified version of the Delta story. More technical details can be found in Seugnet, Simpson, et al. 1997; and Rooke and Xu 1998.

76 *Such polarities also trace back to signaling among cells, involving similar processes to those we have already encountered.* For a technical account of how polarities can be coordinated see Zallen 2007.

78 *When an apple binds to a site, it promotes the binding of another apple to one of the nearby sites—the apple proteins boost each other's attachment.* Such mutual assistance is called cooperative binding and is described for the Bicoid protein in Lebrecht, Foehr, et al. 2005.

78 *This is one of the main reasons that you see a steep rise in the concentration of pears in the middle of the embryo.* Another factor contributing to the distribution of Hunchback is a protein called Nanos at the tail end of the embryo that prevents early copies of the mother's *Hunchback* RNA from being translated into protein. For further details see Sean Carroll's excellent and clear book on developmental diversity (Carroll, Grenier, and Weatherbee 2001).

82 *For example, while mutations play a key part in evolution, there is no exact counterpart to mutation in development.* The nearest equivalents to mutations in development are chemical modifications to DNA, or to proteins intimately bound with DNA, called epigenetic changes (Allis 2009). These modifications may be transmitted from a cell to its daughters. However, epigenetic marks are more reversible than mutations and may be set in a controlled way at certain developmental stages, rather than arising haphazardly.

CHAPTER FOUR

Page

86 *We can continue, adding further colors to our canvas.* Several accessible accounts for the various steps in patterning the fruit fly embryo are available (Coen 1999, Carroll 2005, Nüsslein-Volhard 2006).

92 *Mutations that disrupt the genes coding for the regulatory proteins will result in flies developing with missing segments, or with one type of segment replaced by another.* A readable account of the effect of equivalent mutations in humans has been given by Armand Leroi (Leroi 2003).

92 *The RNA and Bicoid protein then propagate through the egg, forming a gradient from head to tail.* Early studies on the Bicoid gradient concluded that it was formed purely through diffusion of the Bicoid protein from the head to tail end. However, more recent studies show that RNA movement is also involved (Spirov, Fahmy, et al. 2009).

93 *George Stubbs was obsessed with horses.* For an illustrated biography of Stubbs see Morrison 1997.

94 *Jonathan Bard has proposed that the variation between the different species of zebra is related to the time at which patterning happens in the embryo.* See Bard 1977.

101 *As each region tries to grow in a particular way, the computer simulates the way it pulls and pushes against other regions, causing the canvas to deform.* The paper describing the computer model for snapdragon flower development is Green, Kennaway, et al. 2010.

102–3 *Figure 32 illustrates a tree that was generated by a computer program based on this idea.* The paper describing this program is Palubicki, Horel, et al. 2009.

103 *Whether a neuron survives or not can be influenced by whether it receives particular signaling molecules, called neurotrophins.* The story of neurotrophins is given by one of their discoverers, Rita Levi-Montalcini (Levi-Montalcini 2009).

105 *Stubbs published his results in a monumental book,* Anatomy of the Horse, *the best anatomical description of the animal of its time.* Stubb's book is (Stubbs 1776, reprinted 1938).

107–8 *For example, one regulatory protein is present at higher levels toward the central and back end of the cortex (fig. 36,* left*).* More details of neural development are found in the textbook, Sanes, Reh, and Harris 2006.

CHAPTER FIVE

Page

111 *Bashford Dean had two passions in life.* For a short autobiography see Dean 1994. His helmet diagrams are described in Dean 1915.

114 *Yet many of its underlying principles are found in unicellular relatives.* Good accounts of the principles of unicellular biology are to be found in an accessible book by Dennis Bray (Bray 2009), and a more technical but nevertheless clear book by Mark Ptashne and Alexander Gann (Ptashne and Gann 2001).

115 *Most of the ocean is in perpetual darkness, but near its surface there is sufficient light to support a thriving community of microscopic plants.* For an accessible account of the evolution of plants see Corner 1964. More of a textbook account is in Bell 1992.

117 *But there are some advantages for animals being larger, including the ability to swallow other creatures, and avoid being swallowed in turn.* A popular account of how size is important in biology has been given by John Tyler Bonner (Bonner 2006).

118 As Stephen Jay Gould has pointed out, we live in the "age of bacteria," a period which has lasted about 3.5 billion years. See Gould 1996, p. 176.

120 Quotation from Dean 1994, p. 28.

121 *This is because journeys through genetic and developmental space are connected.* For fuller accounts of the relationship between evolution and development see Carroll, Grenier, and Weatherbee 2001; and Stern 2011.

CHAPTER SIX

Page

125 *A good example is the swimming behavior of the bacterium* Escherichia coli. A comprehensive and readable account of how the swimming movements or chemotaxis of microbes is controlled is given in Bray 2009.

128 *Perhaps more surprisingly, plants also have an internal clock.* For an accessible account of biological clocks see Foster and Kreitzman 2005.

129 *Studies on Arabidopsis revealed that its internal clock relies on the daily rise and fall of particular regulatory proteins.* For more details on the way in which plant clocks work and allow day length to be determined see Eriksson and Millar 2003, and Foster and Kreitzman 2010.

132 *If an insect happens to crawl over a leaf of the Venus fly trap (*Dionaea muscipula*), the leaf rapidly snaps shut.* A good, readable book on plant movements is Simons 1992.

132 Darwin quotation on the Venus fly trap is from correspondence with William Canby and is discussed in Jones 1923.

134 *In 1960, when a typhoon hit Tokyo, Hideo Toriyama at the Women's University noticed that after a period of repeated buffeting the local* Mimosa pudica *plants* *stopped closing up their leaves and kept them open.* These findings are described in Toriyama 1966.

135–36 Darwin quotation on the sea slug from Darwin 1890, p. 28.

136 *But things changed in the 1960s, when Eric Kandel at New York University decided to use a Californian variety of this slug,* Aplysia californica *(fig. 45), as a subject for studying the detailed mechanisms of animal responses.* Eric Kandel's discoveries and life story are included in his very readable biography (Kandel 2006).

143–44 *The somatosensory area has a remarkable structure that was revealed by the neurosurgeon Wilder Penfield and colleagues, working at McGill University in Montreal during the 1930s and 1940s.* See Penfield and Rasmussen 1950.

CHAPTER SEVEN

Page

150 *Without making continual predictions, our life would be impossible; we would be in a state of continual anxiety, never knowing what lay round the corner or what to expect.* The importance of prediction in brain function is discussed extensively in a very readable book by Jeff Hawkins and S. Blakeslee (Hawkins and Blakeslee 2004) and in Llinás 2001.

153 Pavlov quotation from Pavlov 1927, p. 13.

154 *This experiment was carried out by Leon Kamin working in Princeton in the 1960s with rats, although I will continue discussing dogs in order to illustrate his results.* Kamin's findings are reported in Kamin 1969.

154–55 *While Pavlov was carrying out his studies in Russia, other scientists like Santiago Ramón y Cajal in Spain and Charles Sherrington in England were looking into the details of how neurons work.* For a readable account of Cajal's work see Rapport 2005. Sherrington's approach and philosophy is described in Sherrington 1951.

155 *In his book* Conditioned Reflexes and Neuron Organisation *(1948), Konorski showed how Pavlov's results could be interpreted as involving specific changes in the formation and number of synaptic connections between neurons.* See Konorski 1948.

155 *Donald Hebb independently arrived at a similar conception in his book* The Organisation of Behaviour *(1949).* See Hebb 1949.

155 *In the mid-1980s, Ranulfo Romo and Wolfram Schultz at the University of Fribourg, Switzerland, were recording electrical signals from particular neurons in a mon-*

key's brain as it learned tasks. See Romo and Schultz 1990. A more popular account has been given by Read Montague (Montague 2006).

157 *Their solution integrates the experimental findings of Romo and Schultz with a mathematical theory of learning, called Temporal Difference learning.* The theory of TD-learning is described in Sutton and Barto 1981, and Sutton 1988; and its neural implementation is discussed in Montague, Dayan, and Sejnowski 1996. For a more popular account of TD-learning see Montague 2006, and for a textbook account see Dayan and Abbott 2005.

162 *For example, we might imagine that the door sound sets off a series of firing patterns of various durations in the brain, some lasting a short time, others a longer time.* This way of representing time, sometimes called the top-hat representation, is described in Suri and Schultz 1999.

163–64 *By contrast, cases like the monkey retrieving the apple, where the action is also learned, are called operant, or instrumental, conditioning.* Conditioning of this kind was first clearly demonstrated in 1928 by Jerzy Konorski and Stefan Miller, who referred to such responses as conditioned reflexes of the second type (classical conditioning was the first type); see Konorski 1948. A similar type of conditioning was independently discovered later by the American psychologist B. F. Skinner, who described it as operant conditioning (Skinner 1953).

164 *We therefore know less about the neural details of punishment as compared to reward learning.* The dopamine-releasing neurons, for example, are stimulated by rewards much more than by aversive treatments (Schultz 2010).

165 *These short-term changes may then build up through further experiences to result in longer term anatomical changes, such as changes in synapse number.* Anatomical changes of this kind are also referred to as structural plasticity (Gogolla, Galimberti, and Coroni 2007).

165 *In 1836, after having just returned from his five-year voyage on the Beagle, Darwin decided to test the memory of his dog.* This story is recounted in Darwin 1901, p. 112.

166 *Darwin had been taught the classics at school, often learning forty or fifty lines from Homer before morning chapel.* See Darwin 1958.

168 *But the same could be said for most mechanisms that have been proposed for learning.* For a technical description of general principles involved in learning mechanisms see Haykin 2009, pp. 396–400. Haykin identifies four principles for self-organization of neural networks: self-amplification (reinforcement), competition, cooperation, and structural information (population variation). The principle of persistence is not explicitly mentioned but is taken for granted.

170 *But the response to meat turns out not to be inborn; it is a result of conditioning.* This experiment, carried out by I. S. Cytovich in 1911 is described in Konorski 1948.

171 *It is through discrepancies that the world announces itself to us.* The importance of discrepancies for our notion of the world was emphasized by the philosopher Martin Heidegger (Heidegger 1927, English translation 1962).

CHAPTER EIGHT

Page

176 Leonardo quotation about bombards is from MacCurdy 1938, Vol. 2, p. 194.

179 *Saccades are not just a feature of human vision; they are also exhibited by many other animals, from squid to goldfish.* See Land 1999.

180 *Part of the answer is that we do not detect visual motion when our eyes are in the process of jumping from one position to the next.* This phenomenon, called saccadic suppression, is described in Bridgeman, Hendry, et al. 1975.

180 *Evidence that we calibrate our eyes through learning comes from the following experiment.* Experiments on saccade adaptation are described in Straube, Fuchs, et al. 1997; and Wallman and Fuchs 1998.

184 *Shown a particular image, our visual-shift neuron fires according to how much that image subsequently moves along on the retina.* This system depends on being able to compare images before and after a saccade, requiring visual information in the brain to persist from one eye fixation to the next; this is known as transsaccadic memory (Irwin 1991).

187 *If someone's eye movements are paralyzed with a small dose of the drug curare, the person will report that the scene in front of them tends to jerk around when they try to move their eyes.* See Stevens, Emerson, et al. 1976.

187 *It is known that there are regions of the brain that are organized as maps that relate visual inputs to eye movements, similar to the sensory and motor maps described in chapter 6.* Some key areas are the frontal eye field and superior colliculus (Purves, Augustine, et al. 2008).

187 *But the story of eye calibration shows how perception, learning, and action are much more interwoven.* An accessible account that emphasizes the feedback between action and perception is by psychologist Ulrich Neisser (Neisser 1976).

189 *Control systems of this kind have the advantage of being able to adapt if the missiles are knocked off course by external factors, such as the wind; they also have the flexibility to follow moving targets.* Control systems and their relevance to living systems have been extensively discussed by William Powers (Forssell 2009).

189 *As before, I want to use our visual system to illustrate the principles involved.* For a clear account of the basis of eye movements see Carpenter 2003; for more technical details see Carpenter 1988.

189 *Like the visual-shift neurons I described earlier, these visual-velocity neurons operate by comparing signals from different photoreceptors of the retina received at various times.* A classic model for how visual velocity might be detected is given in Reichardt 1961, while a more recent study on particular neurons, called starburst amacrine cells, involved in detecting image motion in the retina is described in Euler, Detwiler, and Denk 2002.

190 *In the 1890s, George M. Stratton carried out what has been described by Richard Gregory as perhaps the most famous experiment in all of experimental psychology.* See Stratton 1897, and Gregory 1997.

191 Quotation from Stratton 1897, p. 344.

191 *Discrepancy neurons may be firing because you are getting signals from your visual-velocity neurons, while there is no balancing signal coming from neurons that sense or drive head movements.* Eyes continually jump back to their starting point after following a moving scene for a short time, a process called nystagmus. See Carpenter 2003.

195 *Yet behind the scenes, neurons may be fighting it out to ensure that eventually one action prevails.* For an engaging discussion on the role of neural competition in guiding our actions, see Eagleman 2011.

195 *The monkey is learning not only about rewards in relation to individual actions, but also the best policy to adopt for getting a reward given a range of choices and situations.* For discussion of policies and how they may be learned see Daw, Niv, and Dayan 2005; and Dayan and Abbott 2005.

195 *Suppose, for example, that the monkey gets fed up with eating lots of apple and yearns for a new food, such as banana.* The interplay of incentives and rewards in instrumental learning is discussed in Balleine and Dickinson 1998.

196 *The main difference is that we have introduced further levels of interaction between actions and their effects; further degrees of competition, cooperation, combinatorial richness, and recurrence.* For a nontechnical account of decision making see Montague 2006; a more technical account is in the later chapters of Dayan and Abbott 2005.

196–97 Helmholtz quotation from Helmholtz 1866, pp. 30–31.

197 *In the 1990s, Vittorio Gallese, Giacomo Rizzolatti, and colleagues at the University of Parma were recording electrical signals from neurons in the brains of macaque monkeys.* See Gallese, Fadiga, et al. 1996. For an account of mirror neurons in humans see Mukamel, Ekstrom, et al. 2010, and for a more general discussion of their possible significance for human interactions see Ramachnandran 2011.

198 Chris Frith quoted from his clear and accessible book on the mind (Frith 2007, p. 175).

CHAPTER NINE

Page

200 *On a more modest scale, we perform similar creative acts every day, by activities like arranging flowers in a vase or constructing a new sentence when we speak.* An interesting discussion of how everyday statements involve seeing things in particular ways is given by George Lakoff and Mark Johnson (Lakoff and Johnson 1980).

202 *For example, there is an area known as the fusiform face area that is involved in recognition of complex objects like faces.* For an accessible book on face recognition see Bruce and Young 2000.

202 *People with lesions in this area have difficulty in interpreting or recognizing faces, a condition known as prosopagnosia.* For some examples of people with prosopagnosia see the very readable account by Oliver Sacks in *The Man Who Mistook his Wife for a Hat* (Sacks 1985); see also Etcoff, Freeman, et al. 1991.

204 *This can happen through a process called presynaptic inhibition, in which the firing of one neuron can interfere with the ability of another to release neurotransmitters from its terminal.* Although I use presynaptic inhibition to illustrate attention shifts, any mechanism which preferentially blocks or enhances the signaling of a neuron could be employed to the same effect. For a technical account of the biophysical principles of neural computation see Koch 1999.

208 *As pointed out by David Van Essen and colleagues at Caltech in 1993, multiple neural eyes could help with making visual interpretations.* See Olshausen, Anderson, and Van Essen 1993.

210 *The neural signals from this original view of Mary at the center form a standard reference that the neural system remembers and always tries to match.* Learning the neural signals associated with the original or averaged view of Mary could be established through discrepancy neurons similar to those described for figure 61. The discrepancy neurons would come to balance signals from the neural eyes against a fixed uniform neural input. If Mary is no longer in view, the discrepancy neurons then fire in a compensatory pattern that matches Mary's signals, providing a memory or reference for Mary's appearance.

210 *We need not go into the details of how this happens as it involves neural learning circuits and principles similar to those we already encountered for discrepancy neurons.* The principles of correlative learning, using Hebbian neural networks (named after psychologist Donald Hebb) are described in several technical accounts (Diamantaras and Kung 1996, and Haykin 2009).

212 *In practice, objects do not show such erratic behavior, so we are able to learn what are called invariant representations—models that allow us to perceive an unvarying identity when we are looking at something.* The construction of invariant representations based on the slow rate at which objects change when viewed continuously can potentially be achieved through several neural learning mechanisms, such as slow feature analysis (Wiskott and Sejnowski 2002); this has recently received experimental support (Li and DiCarlo 2008).

214 *Visual information is indeed dealt with at many levels in the human brain.* For a good and readable account of the various levels of analysis in the visual cortex see Zeki 1993.

216 *Something like this happens in people suffering from* visual form agnosia. See Milner and Goodale 1996.

216–17 *Changes detected in our neural eyes at levels 2 and 3 are then fed back to connections with level 1, influencing where the neural eyes are looking.* Connections in the brain that run from higher to lower levels of the visual cortex are well documented and known as "reentrant" (Zeki 1993).

221 *For an indication of this average, look at figure 73. This image was generated by my placing positions for neural eyes on 179 portraits by artists such as Rembrandt, Modigliani, and Leonardo.* The software used to generate this average was developed by my colleagues Andrew Bangham, Barry Theobald, and Andy Hanna using image alignment and warping methods.

222 *This process of neural correlation and distillation can be simulated with a computer.* The procedure used to identify the various trends is called principal component analysis, which has been shown to be equivalent to learning through Hebbian neural networks (Diamantaras and Kung 1996, and Haykin 2009). The application used here employs active shape models and image synthesis (Cootes, Taylor, et al. 1995; Vetter and Poggio 1997) using software developed by my colleagues Andrew Bangham, Barry Theobald, and Andy Hanna.

226 *Caricaturists are experts at manipulating our nuerons.* For further discussion of caricatures see the readable book by Bruce and Young 2000.

228 *[Cézanne] may instinctively feel compelled to add some color here or there, or feel that a brush stroke works well or is not quite right.* A good illustration of the instinctive dialogue between artist and canvas comes from artist Lucien Freud's comments, made while painting a portrait. According to one of his subjects Martin Gayford: "When he is really concentrating he mutters constantly, giving himself instructions: 'Yes, perhaps—a bit,' 'Quite!', 'No-o, I don't think so,' 'A bit more yellow' " (Gayford 2010, p. 43).

230 *A new explanation may seem to suddenly occur to a scientist, just as a poet may unexpectedly think of a lyrical line.* For a discussion on the issue of how major discoveries arise see Robinson 2010. See also David Eagleman's very readable and clear account of the role of the unconscious in our thoughts and actions (Eagleman 2011).

230 Darwin quotation from Darwin 1958, p. 43.

CHAPTER TEN

Page

234 *In the 1960s, David Hubel and Torsten Wiesel working at the Harvard Medical School in Boston, started to address this question by make detailed recordings from the visual cortex of cats and monkeys.* These experiments are described in a compilation of their papers (Hubel and Wiesel 2005).

236 *Later computational studies showed that the experimental observations of Hubel and Wiesel could be accounted for by mechanisms similar to those invoked by Alan Turing to explain spotty or stripy patterns (chapter 3).* See Von der Malsburg 1979; Swindale 1980; and Miller, Keller, and Stryker 1989.

240 *Psychologists Michael Cook at the University of Wisconsin and Susan Mineka at Northwestern University explored this further by showing infant monkeys movie clips of adults looking fearful, followed each time by clips of various objects.* See Cook and Mineka 1989.

243 *Although more indirect than instinctive responses, learning provides greater flexibility for coping with the complexities and contingencies of what the animal encounters.* The relationship between instinct and learning is nicely discussed in Eagleman 2011. The argument based on flexibility may apply even for predictable factors in the environment, because of greater efficiencies in finding adaptive solutions brought about through learning, as emphasized by Geoff Hinton (Hinton and Nowlan 1987).

CHAPTER ELEVEN

Page

245 *The historians J. R. and William McNeill, for example, have emphasized how competition and cooperation have been critical for driving human cultural change.* See McNeill and McNeill 2003.

245 *In the middle of the fifteenth century, the goldsmiths Andrea Verrocchio and Antonio Pollaiuolo were fighting it out in Florence.* For a good account see Brown 1998. Some readable biographies of Leonardo are Nicholl 2004 and Kemp 2004.

248 *This population and its nearby towns produced several great artists of the Renaissance, including Leonardo, Michelangelo, and Botticelli.* Michelangelo was born in 1475 in Caprese, about forty miles from Florence.

249 *These modern equivalents do not produce masterpieces like the* Mona Lisa *or* God Creating Adam *because they were born and brought up in a very different setting.* For an accessible and entertaining account of the importance of context in achievement see Gladwell 2009.

250 *Instead of isolated populations, the world comprised many interconnected ones.* These populations form what has been called the old world web, nicely and accessibly described by J. R. and William McNeill (McNeill and McNeill 2003).

251 Reynolds quotation from Reynolds 1992 (first published in 1769), p. 94.

251 *By 1500, Gutenberg-style presses could be found in 236 towns in Europe, and had printed about twenty million books in various languages.* Figures taken from McNeill and McNeill 2003; see also Eisenstein 1983.

253 The quotation from Leonardo's letter comes from Nicholl 2004, p. 180.

254 *Similarly, the success and fame that Leonardo was to later achieve stimulated the achievements of competitors, such as Michelangelo.* For a popular account of the competition between Leonardo and Michelangelo see Jones 2010.

256 *And without the ability to communicate, you would be unable to benefit from others, and your achievements would be unlikely to persist and spread.* For a study of the way people may benefit from each other in making decisions see Bahrami, Olsen, et al. 2010.

256 *Similarly, our propensity to exchange money and goods is a cooperative venture.* The importance of exchange and commerce for cultural change has been emphasized by Matt Ridley in his book *The Rational Optimist* (Ridley 2010).

256 Leonardo's quotation about sculpture comes from McCurdy 1938, Vol. 2, p. 214.

256 Byron quotation taken from Prothero, ed. 1966, Vol. 4, p. 107.

257 Rutherford quotation is from Blackett 1955, p. 19.

257 Leonardo's quotation about the benefits of working together is from MacCurdy 1938, Vol. 2, p. 241.

258 Reynolds quotation from Reynolds 1992 (first published 1769), p. 91.

258 *Over time, our species has journeyed through cultural space, taking a particular path as human habits and endeavors have become modified across the globe.* For readable accounts of the forces influencing the course of human culture see books by Jared Diamond (Diamond 1998) and Neil MacGregor (MacGregor 2010).

259 *The earliest human artifacts, stone chopping tools from Africa, date back to around two million years ago.* For nicely illustrated accounts of human evolution and early tools see Palmer 2010 and MacGregor 2010.

259 *Developments in language and communication were no doubt involved, but exactly how* Homo sapiens *became the dominant human species is a subject of much discussion.* For a comprehensive review of how human language may have evolved see Fitch 2010, and for a discussion on the neural basis of language acquisition see Ramachnandran 2011.

261 Quotation from Leonardo on criteria for painting is from MacCurdy 1938, p. 265.

264 *For example, a unit of transmission called the meme has been proposed as playing a similar role in culture as a gene plays in evolution.* Memes were first defined by Richard Dawkins in Dawkins 1976. A further description of memes and their application is in Blackmore 1999.

264 Quotation on memes comes from Coyne 1999, p. 768.

CHAPTER TWELVE

Page

268 *For a given set of options, we therefore tend to seek the one that best matches our values: the best food, the best partner, the best artifact, the best means of communicating, the best way of getting around, or the best explanation.* For an account of how we are continually driven to seek rewards see Berns 2005.

270 *We can't readily imagine four-dimensional objects.* The evolutionary biologist J. B. S. Haldane also explores extra dimensions and different worlds in his essay *Possible Worlds* (Haldane 1927). Haldane tries to envisage a fourth dimension through geometry rather than color.

272–73 *As Einstein realized, the theory starts to break down when bodies approach the speed of light, and it fails to account correctly for the effect of gravity on light itself.* A good account of relativity and its relation to quantum mechanics is given by Brian Greene (Greene 2004).

273 Quotation is from Rees 1999, p. 37.

274 *Examples like rainbow world, relativity, and quantum mechanics highlight what many philosophers have also pointed out: we cannot access our world directly, but are always forced to look at it through particular frameworks.* Perhaps the philosopher best known for exploring this issue in depth is Immanuel Kant (Kant 1781 (English translation, 1986)).

275 *He developed a terrible fear of being touched, as Émile Bernard came to realize after he made the mistake of trying to help him from falling one day.* This story is told in Doran 2001, p. 70.

275 Geffroy quotation is from Doran 2001, p. 6.

275 Cézanne quotation is from Doran 2001, p. 18.

References

Acker, W.B.R. (1954). *Some T'ang and Pre-T'ang Texts on Chinese Paintings*. Dover, New York.

Allis, C. D. (2009). *Epigenetics*. Cold Spring Harbor Laboratory Press, Cold Spring Harbor, N.Y.

Bahrami, B., K. Olsen, et al. (2010). "Optimally Interacting Minds." *Science* 329(5995): 1081–5.

Ball, P. (2000). *H2O: A Biography of Water*. Phoenix Press, London.

Balleine, B. W. and A. Dickinson (1998). "Goal-Directed Instrumental Action: Contingency and Incentive Learning and Their Cortical Structures." *Neuropharmacology* 37: 407–19.

Bard, J. (1977). "A Unity Underlying the Different Zebra Striping Patterns." *Journal of Zoology* 183: 527–39.

Barton, N. H., D.E.G. Briggs, et al. (2007). *Evolution*. Cold Spring Harbor Laboratory Press, Cold Spring Harbor, N.Y.

Bell, P. R., A. R. Hemsley (1992). *Green Plants: Their Origin and Diversity*. Cambridge University Press, Cambridge.

Berns, G. (2005). *Satisfaction: The Science of Finding True Fulfillment*. Henry Holt, New York.

Blackett, P.M.S. (1955). "Rutherford Memorial Lecture 1954." *Physical Society Yearbook*: 13–22.

Blackmore, S. (1999). *The Meme Machine*. Oxford University Press, Oxford.

Bonner, J. T. (2006). *Why Size Matters: From Bacteria to Blue Whales*. Princeton University Press, Princeton, N. J.

Boring, E. G. (1930). "A New Ambiguous Figure." *American Journal of Psychology* 42: 444–45.

Bray, D. (2009). *Wetware: A Computer in Every Living Cell*. Yale University Press, New Haven, Conn.

Bridgeman, B., D. Hendry, et al. (1975). "Failure to Detect Displacement of the Visual World During Saccadic Eye Movements." *Vision Research* 15: 719–22.

Brown, D. A. (1998). *Leonardo da Vinci: Origins of a Genius*. Yale University Press, New Haven, Conn.

Bruce, V. and A. Young (2000). *In the Eye of the Beholder: The Science of Face Perception*. Oxford University Press, Oxford.

Carpenter, R.H.S. (1988). *Movements of the Eyes*. Pion, London.

Carpenter, R.H.S. (2003). *Neurophysiology*. Edward Arnold, London.

Carroll, S. B. (2005). *Endless Forms Most Beautiful: The New Science of Evo Devo and the Making of the Animal Kingdom*. W. W. Norton, New York.

Carroll, S. B., J. K. Grenier, and S. Weatherbee (2001). *From DNA to Diversity: Molecular Genetics and the Evolution of Animal Design*. Blackwell Science, Malden, Mass.

Coen, E. (1999). *The Art of Genes*. Oxford University Press, Oxford.

Cook, M. and S. Mineka (1989). "Observational Conditioning of Fear to Fear-Relevant Versus Fear-Irrelevant Stimuli in Rhesus Monkeys." *Journal of Abnormal Psychology* 98(4): 448–59.

Cootes, T. F., C. J. Taylor, et al. (1995). "Active Shape Models—Their Training and Application." *Computer Vision and Image Understanding* 61(1): 38–59.

Corner, E. H. (1964). *The Life of Plants*. Weidenfeld and Nicolson, London.

Coyne, J. A. (1999). "The Self-Centred Meme." *Nature* 398: 767–68.

Crick, F. (1989). "Neural Edelmanism." *Trends in Neurosciences* 12: 240–48.

Darwin, C. (1890). *A Journal of Researches into the Natural History and Geology of the Countries Visited during the Voyage of H.M.S. "Beagle" Round the World*. Ward Lock, London.

Darwin, C. (1958). *The Autobiography of Charles Darwin and Selected Letters*, ed. Francis Darwin. Dover, New York.

Darwin, C. (1901). *The Descent of Man and Selection in Relation to Sex*, 2d ed. John Murray, London.

Daw, N. D., Y. Niv, and P. Dayan (2005). "Uncertainty-Based Competition Between Prefrontal and Dorsolateral Striatal Systems for Behavioral Control." *Nature Neuroscience* 8(12): 1704–11.

Dawkins, R. (1976). *The Selfish Gene*. Oxford University Press, Oxford.

Dawkins, R. (1986). *The Blind Watchmaker*. Penguin, London and New York.

Dayan, P. and L. F. Abbott (2005). *Theoretical Neuroscience: Computational and Mathematical Modeling of Neural Systems*. MIT Press, Cambridge, Mass.

Dean, B. (1915). "An Explanatory Label for Helmets." *Bulletin of the Metropolitan Museum of Art* 10: 173–77.

Dean, B. (1994). "The Hobby of Collecting Ancient Armor." *American Society of Arms Collectors Bulletin* 70: 24–28.

Dennett, D. C. (1996). *Darwin's Dangerous Idea: Evolution and the Meanings of Life*. Penguin, London.

Diamantaras, K. I. and S. Y. Kung (1996). *Prinicpal Component Neural Networks: Theory and Applications*. Wiley, New York.

Diamond, J. (1998). *Guns, Germs and Steel: A Short History of Everybody for the Last 13,000 Years*. Vintage, London.

Doran, M. (2001). *Conversations with Cézanne*. University of California Press, Berkeley, Calif.

Eagleman, D. (2011). *Incognito: The Secret Lives of the Brain*. Cannongate, Edinburgh.

Edelman, G. (1994). *Bright Air, Brilliant Fire: On the Matter of Mind*. Penguin, London.

Eisenstein, E. L. (1983). *The Printing Revolution in Early Modern Europe*. Cambridge University Press, Cambridge.

Eriksson, M. E. and A. J. Millar (2003). "The Circadian Clock. A Plant's Best Friend in a Spinning World." *Plant Physiology* 132: 732–38.

Etcoff, N. L., R. Freeman, and K. R. Cave (1991). "Can We Lose Memories of Faces? Content Specificity and Awareness in a Prosopagnosic." *Journal of Cognitive Neuroscience* 3: 25–41.

Euler, T., P. B. Detwiler, and W. Denk (2002). "Directionally Selective Calcium Signals in Dendrites of Starburst Amacrine Cells." *Nature* 418(6900): 845–52.

Fitch, W.T. (2010). *The Evolution of Language*. Cambridge University Press, Cambridge.

Forssell, D.C., ed (2009). *Perceptual Control Theory: Science and Applications—A Book of Readings*. Living Control Systems Publishing: www.livingcontrolssystems.com.

Foster, R. and L. Kreitzman (2010). *Seasons of Life: The Biological Rhythms That Enable Living Things to Thrive and Survive*. Profile Books, London.

Foster, R. G. and L. Kreitzman (2005). *Rhythms of Life: The Biological Clocks that Control the Daily Lives of Every Living Thing*. Profile Books, London.

Frith, C. (2007). *Making up the Mind: How the Brain Creates our Mental World*. Blackwell, Malden, Mass.

Gallese, V., L. Fadiga, et al. (1996). "Action Recognition in the Premotor Cortex." *Brain* 119: 593–609.

Gavrilets, S. (2004). *Fitness Landscapes and the Origin of Species*. Princeton University Press, Princeton, N.J.

Gayford, M. (2010). *Man with a Blue Scarf: On Sitting for a Portrait by Lucian Freud*. Thames and Hudson, London.

Gladwell, M. (2009). *Outliers: The Story of Success*. Penguin, London and New York.

Gogolla, N., I. Galimberti, and P. Coroni (2007). "Structural Plasticity of Axon Terminals in the Adult." *Current Opinion in Neurobiology* 17(5): 516–24.

Gould, S. J. (1977). *Ontogeny and Phylogeny*. Harvard University Press, Cambridge Mass.

Gould, S. J. (1996). *Life's Grandeur: The Spread of Excellence from Plato to Darwin*. Jonathan Cape, London.

Green, A. A., J. R. Kennaway, et al. (2010). "Genetic Control of Organ Shape and Tissue Polarity." *PLoS Biology* 8(11): e1000537.

Greene, B. (2004). *The Fabric of the Cosmos*. Alfred A. Knopf, New York.

Gregory, R. (1997). *Mirrors in Mind*. W. H. Freeman Spektrum, Oxford.

Haldane, J.B.S. (1927). *Possible Worlds and Other Essays*. Chatto and Windus, London.

Hawkins, J. and S. Blakeslee (2004). *On Intelligence*. Henry Holt, New York.

Haykin, S. (2009). *Neural Networks and Learning Machines*. Pearson Prentice Hall, Upper Saddle River, N. J.

Hebb, D. O. (1949). *The Organisation of Behaviour: a Neuropsychological Theory*. Wiley, New York.

Heidegger, M. (1962). *Being and Time*. Trans. J. Macquarrie and E. Robinson. Basil Blackwell, Oxford.

Helmholtz, H. (1924). *Helmholtz's Treatise on Physiological Optics*. Vol. 3 Trans. J.P.C. Southall. Optical Society of America, New York. First published in 1866 as *Handbuch der Physiologischen Optik*.

Hinton, G. E. and S. J. Nowlan (1987). "How Learning Can Guide Evolution." *Complex Systems* 1: 495–502.

Hodges, A. (1992). *Alan Turing: The Enigma*. Vintage, London.

Howard, J. (1982). *Darwin*. Oxford University Press, Oxford.

Hubbell, S. P. (2001). *The Unified Neutral Theory of Biodiversity and Biogeography*. Princeton University Press, Princeton, N. J.

Hubel, D. H. and T. N. Wiesel (2005). *Brain and Visual Perception: The Story of a 25-Year Collaboration*. Oxford University Press, Oxford.

Irwin, D. E. (1991). "Information Integration Across Saccadic Eye Movements." *Cognitive Psychology* 23: 420–56.

Jones, F. M. (1923). "The Most Wonderful Plant in the World." *Natural History* 23(6): 589–96.

Jones, J. (2010). *The Lost Battles: Leonardo, Michelangelo and the Artistic Duel that Defined the Renaissance*. Simon and Schuster, London.

Juniper, B. E. and D. J. Mabberley (2006). *The Story of the Apple*. Timber Press, Portland, Oreg.

Juniper, B. E., R. J. Robins, and D. M. Joel (1989). *The Carnivorous Plants*. Academic Press, London.

Kamin, L. J. (1969). "Predictability, Surprise, Attention and Conditioning." In *Punishment and Aversive Behavior*, ed. B. A. Campbell and R. M. Church. Appleton-Century-Crofts, New York, pp. 279–96.

Kandel, E. R. (2006). *In Search of Memory: The Emergence of a New Science of Mind*. W. W. Norton, New York.

Kant, I. (1986). *Critique of Pure Reason*. Trans. J. M. D. Meiklejohn. Dent, London and Melbourne. First published in 1781.

Kemp, M. (2004). *Leonardo*. Oxford University Press, Oxford.

Knoll, A. H. (2003). *Life on a Young Planet: The First Three Billion Years of Evolution on Earth*. Princeton University Press, Princeton, N. J.

Koch, C. (1999). *Biophysics of Computation*. Oxford University Press, New York.

Kondo, S. and T. Miura (2010). "Reaction-Diffusion Model as a Framework for Understanding Biological Pattern Formation." *Science* 329(5999): 1616–20.

Konorski, J. (1948). *Conditioned Reflexes and Neuron Organisation*. Cambridge University Press, Cambridge.

Kruse, K., M. Howard, and W. Margolin (2007). "An Experimentalist's Guide to Computational Modelling of the Min System." *Molecular Microbiology* 63(5): 1279–84.

Lakoff, G. and M. Johnson (1980). *Metaphors We Live By*. University of Chicago Press, Chicago, Ill.

Land, M. F. (1999). "Motion and Vision: Why Animals Move Their Eyes." *Journal of Comparative Physiology A* 185: 341–52.

Lane, N. (2010). *Life Ascending: The Ten Great Inventions of Evolution*. Profile, London.

Lebrecht, D., M. Foehr, et al. (2005). "Bicoid Cooperative DNA Binding is Critical for Embryonic Patterning in Drosophila." *Proceedings of the National Academy of Sciences of the United States of America* 102(37): 13176–81.

Leroi, A. M. (2003). *Mutants: On the Form, Varieties and Errors of the Human Body*. Harper Collins, London.

Levi-Montalcini, R. (2009). "Nerve Growth Factor 35 Years Later." *Science* 237: 1154–62.

Li, N. and J. J. DiCarlo (2008). "Unsupervised Natural Experience Rapidly Alters Invariant Object Representation in Visual Cortex." *Science* 321(5895): 1502–507.

Llinás, R., R. (2001). *I of the Vortex*. MIT Press, Cambridge Mass.

MacCurdy, E. (1938). *The Notebooks of Leonardo da Vinci*. Reprint Society, London.

MacGregor, N. (2010). *A History of the World in 100 Objects*. Allen Lane, London.

Mayr, E. (2002). *What Evolution Is*. Phoenix Press, London.

McNeill, J. R. and W. H. McNeill (2003). *The Human Web: A Bird's-Eye View of World History*. W. W. Norton, New York and London.

Meinhardt, H. (1998). *The Algorithmic Beauty of Sea Shells*. Springer-Verlag, Berlin.

Miller, K. D., J. B. Keller, and M. P. Stryker (1989). "Ocular Dominance Column Development: Analysis and Simulation." *Science* 245: 605–15.

Milner, A. D. and M. A. Goodale (1996). *The Visual Brain in Action*. Oxford University Press, Oxford.

Montague, P. R., P. Dayan, and T. J. Sejnowski (1996). "A Framework for Mesencephalic Dopamine Systems Based on Predictive Hebbian Learning." *Journal of Neuroscience* 16(5): 1936–47.

Montague, R. (2007). *Your Brain Is (Almost) Perfect: How we Make Decisions*. Plume, New York.

Morrison, V. (1997). *The Art of George Stubbs*. Grange Books, London.

Mukamel, R., A. D. Ekstrom, et al. (2010). "Single-Neuron Responses in Humans during Execution and Observation of Actions." *Current Biology* 20: 750–56.

Murray, H.J.R. (1913). *A History of Chess*. Oxford University Press, Oxford.

Neisser, U. (1976). *Cognition and Reality*. W. H. Freeman, San Francisco, Calif.

Nicholl, C. (2004). *Leonardo da Vinci: The Flights of the Mind*. Penguin, London.

Noble, D. (2006). *The Music of Life: Biology beyond the Genome*. Oxford University Press, Oxford.

Nüsslein-Volhard, C. (2006). *Coming to Life: How Genes Drive Development*. Yale University Press, New Haven, Conn..

Olshausen, B. A., C. H. Anderson, and D. C. Van Essen (1993). "A Neurobiological Model of Visual Attention and Invariant Pattern Recognition Based on Dynamic Routing of Information." *Journal of Neuroscience* 13(11): 4700–719.

Palmer, D. (2010). *Origins: Human Evolution Revealed*. Mitchell Beazley, London.
Palubicki, W., K. Horel, et al. (2009). "Self-Organizing Tree Models for Image Synthesis." *ACM Transactions on Graphics* 28(3): 1–58.
Pavlov, I. P. (1927). *Conditioned Reflexes: An Investigation of the Physiological Activity of the Cerebral Cortex*. Trans. and ed. G. V. Anrep. Oxford University Press, Oxford.
Penfield, W. and T. Rasmussen (1950). *The Cerebral Cortex of Man*. MacMillan, New York.
Pinker, S. (1994). *The Language Instinct*. Penguin, London.
Prothero, R. E., ed. (1966). *The Works of Lord Byron. Letters and Journals*. Octagon Books, New York.
Ptashne, M. and A. Gann (2001). *Genes and Signals*. Cold Spring Harbor Laboratory Press, Cold Spring Harbor, N.Y.
Purves, D., G. L. Augustine, et al. (2008). *Neuroscience*. Sinauer, Sunderland, Mass., US.
Ramachnandran, V. S. (2011). *The Tell-Tale Brain: Unlocking the Mystery of Human Nature*. William Heinemann, London.
Rapport, R. (2005). *Nerve Endings: The Discovery of the Synapse*. W. W. Norton, New York.
Rees, M. (1999). *Just Six Numbers: The Deep Forces that Shape the Universe*. Phoenix Press, London.
Reichardt, W. (1961). "Autocorrelation, a Principle for the Evaluation of Sensory Information by the Central Nervous System." In *Principles of Sensory Communication*, ed. W. A. Rosenblith. John Wiley, New York, pp. 303–17.
Reynolds, J. (1992). *Discourses*. Penguin, London. First published in 1769.
Ridley, M. (1993). *The Red Queen: Sex and the Evolution of Human Nature*. Viking, London.
Ridley, M. (2010). *The Rational Optimist: How Prosperity Evolves*. Fourth Estate, London.
Robinson, A. (2010). *Sudden Genius? The Gradual Path to Creative Breakthroughs*. Oxford University Press, Oxford.
Romo, R. and W. Schultz (1990). "Dopamine Neurons of the Monkey Midbrain: Contingencies of Responses to Active Touch During Self-Initiated Arm Movements." *Journal of Neurophysiology* 63(3): 592–606.
Rooke, J. E. and T. Xu (1998). "Positive and Negative Signals Between Interacting Cells for Establishing Neural Fate." *Bioessays* 20(3): 209–14.
Sachs, J. (1887). *Lectures on the Physiology of Plants*. Oxford University Press, Oxford.
Sacks, O. (1986). *The Man Who Mistook his Wife for a Hat*. Picador, London.
Sacks, O. (1995). *An Anthropologist on Mars*. Picador, London.
Sanes, D. H., T. A. Reh, and W. A. Harris (2006). *Development of the Nervous System*. Elsevier, Burlington, Mass.
Schultz, W. (2010). "Dopamine Signals for Reward Value and Risk: Basic and Recent Data." *Behavioral and Brain Functions* 6: article 24: 1–9 (online).

Seugnet, L., P. Simpson, et al. (1997). "Transcriptional Regulation of Notch and Delta: Requirement for Neuroblast Segregation in Drosophila." *Development* 124(10): 2015–25.

Shenk, D. (2007). *The Immortal Game: A History of Chess or How 32 Carved Pieces on a Board Illuminated Our Understanding of War, Art, Science, and the Human Brain*. Souvenir Press, London.

Sherrington, C. (1951). *Man on His Nature*. Cambridge University Press, Cambridge.

Simons, P. (1992). *The Action Plant*. Blackwell, Oxford and Cambridge.

Skinner, B. F. (1953). *Science and Human Behavior.* Macmillan, New York.

Smith, J. M. and E. Szathmáry (1995). *The Major Transitions in Evolution*. W. H. Freeman, Oxford.

Smith, P. (1996). *Interpreting Cézanne*. Tate Publishing, London.

Spirov, A., K. Fahmy, et al. (2009). "Formation of the Bicoid Morphogen Gradient: An mRNA Gradient Dictates the Protein Gradient." *Development* 136(4): 605–14.

Stern, D. (2011). *Evolution, Development, and the Predictable Genome*. Roberts, Greenwood Village, Colo.

Stevens, J. K., R. C. Emerson, et al. (1976). "Paralysis of the Awake Human: Visual Perceptions." *Vision Research* 16: 93–98.

Stratton, G. M. (1897). "Vision without Inversion of the Retinal Image." *Psychological Review* 4: 341–60.

Straube, A., A. F. Fuchs, et al. (1997). "Characteristics of Saccadic Gain Adaptation in Rhesus Macaques." *Journal of Neurophysiology* 77: 874–895.

Stubbs, G. (1938). *The Anatomy of the Horse*. G. Heywood Hill, London. Originally printed for the author by J. Purser in 1766.

Suri, R. E. and W. Schultz (1999). "A Neural Network Model with Dopamine-Like Reinforcement Signal That Learns a Spatial Delayed Response Task." *Neuroscience* 91(3): 871–90.

Sutton, R. (1988). "Learning to Predict by the Methods of Temporal Differences." *Machine Learning* 3: 9–44.

Sutton, R. S. and A. G. Barto (1981). "Toward a Modern Theory of Adaptive Networks: Expectation and Prediction." *Psychological Review* 88: 135–71.

Swindale, N. V. (1980). "A Model for the Formation of Ocular Dominance Stripes." *Proceedings of the Royal Society of London*. Series B. 208: 243–264.

Toriyama, H. (1966). "The Behaviour of the Sensitive Plant in a Typhoon." *Botanical Magazine* 79: 427–8.

Turing, A. (1952). "The Chemical Basis of Morphogenesis." *Philosophical Transactions of the Royal Society of London*. Series B. 237: 37–72.

Vetter, T. and T. Poggio (1997). "Linear Object Classes and Image Synthesis From a Single Example Image." *IEEE Transactions on Pattern Analysis and Machine Intelligence* 19(7): 733–42.

Von der Malsburg, C. (1979). "Development of Ocularity Domains and Growth Behaviour of Axon Terminals." *Biological Cybernetics* 32: 49–62.

Von der Malsburg, C. (1990). "Network Self-Organisation." In *An Introduction to Neural and Electronic Networks,* ed. S. F. Zornetzer, J. L. Davis and C. Lau. Academic Press, San Diego, Calif., pp. 421–32.

Wallman, J. and A. Fuchs (1998). "Saccadic Gain Modification: Visual Error Drives Motor Adaptation." *Journal of Neurophysiology* 80: 2405–16.

Wiskott, L. and T. J. Sejnowski (2002). "Slow Feature Analysis: Unsupervised Learning of Invariances." *Neural Computation* 14: 715–70.

Wolpert, L. and C. Tickle (2011). *Principles of Development*. Oxford University Press, Oxford.

Zallen, J. A. (2007). "Planar Polarity and Tissue Morphogenesis." *Cell* 129(6): 1051–63.

Zeki, S. (1993). *A Vision of the Brain*. Blackwell Scientific, Oxford.

Illustration Credits

Figure 5 / Plate 1. *Portrait of Ambroise Vollard*, 1899 (oil on canvas), by Paul Cézanne (1839–1906). Musée de la Ville de Paris, Musée du Petit-Palais, Paris, France/ Giraudon/ The Bridgeman Art Library. 35

Figure 8. *God Creating Adam*, 1510, by Michelangelo Buonarroti. Sistine Chapel, Vatican, Rome. 48

Figure 9 / Plate 3. *The Rue Montorgueil, Paris, Celebration of June 30, 1878* (oil on canvas), 1878, by Claude Monet (1840–1926). Musée d'Orsay, Paris, France/ Giraudon/ The Bridgeman Art Library. 49

Figure 10. *The Tower of Babel*, 1563, by Pieter Bruegel the Elder. Kunsthistorisches Museum, Vienna. 54

Figure 11. Pattern obtained by applying Turing's rules. Courtesy of Stan Marée. 63

Figure 24 / Plate 4. *The Garden at Les Lauves*, 1906, by Paul Cézanne. The Phillips Collection, Washington DC. 85

Figure 26. Interactions between themes and variations in a fruit fly embryo. After Coen 1999, p. 300, fig. 15.18. 91

Figure 27. *Zebra*, 1763 by George Stubbs (1724–1806). Yale Center for British Art, Paul Mellon Collection. 95

Figure 28. Stripes generated in different types of zebra. Carroll (2005), figure 9.8, 241. Courtesy of Sean Carroll. 96

Figure 29 / Plate 5. *Apples and Biscuits*, c. 1880, by Paul Cézanne. Musée National de l'Orangerie, Paris. Jean Walter and Paul Guillaume collection. Deformed version courtesy of Richard Kennaway. 98

Figure 30 / Plate 6. Deformed version of Cézanne's *Apples and Biscuits* courtesy of Richard Kennaway. 100

Figure 32. Tree generated from a computer program. From Palubicki, W., K. Horel, et al. (2009). 103

Figure 33. Computer simulated shape changes in *Dictyostelium*. Courtesy of Stan Marée. 104

Figure 34. *Anatomy of the Horse,* 1766, by George Stubbs. Thirteenth Anatomical Table. Royal Academy of Arts, London. 106

Figure 36. Developing cerebral cortex. After Sanes et al., 2006, p. 53. 108

Figure 37. Helmet evolution. From Dean 1915. The Metropolitan Museum of Art. Diagram by Bashford Dean, Department of Arms and Armor. Image © The Metropolitan Museum of Art. 112

Figure 38. Bladder wrack (*Fucus vesiculosus*). After Bell 1992, p. 92, Fig. 4.19. 119

Figure 39. *Large Piece of Turf*, 1503, by Albrecht Dürer. Albertina, Vienna. 123

Figure 41 / Plate 7. *The Librarian*, 1565, by Giuseppe Arcimboldo. Eva Hökenberg Collection, Stockholm. 128

Figure 42 / Plate 8. The seasons by Giuseppe Arcimboldo (1527–1593): *Spring*, *Summer*, *Autumn*, and *Winter*, 1573 (oil on canvas). Louvre, Paris, France/ Giraudon/ The Bridgeman Art Library. 129

Figure 43. Leaf of Venus fly trap. From Juniper et al., 1989, p. 52. 133

Figure 44. Leaves of *Mimosa pudica*. From Sachs, 1887, p. 645, fig. 373. 134

Figure 46. Two neurons. After Rains 2001, p. 32. 137

Figure 47. Neurons in abdominal ganglion of the sea slug *Aplysia californica*. After Kandel 2006, p. 224. 138

Figure 48. The brain of *Aplysia*. After Kandel 2006, p. 147. 139

Figure 50. The human nervous system. From Purves et al., 2008, p. 15, fig. 1.1. 144

Figure 51. Somatosensory and motor area. After Penfield and Rasmussen, 1950. 145

Figure 52. *Libyan Sibyl*, 1511, by Michelangelo Buonarotti. Sistine Chapel, Vatican, Rome. 151

Figure 57. Drawing of bombards on folio 33r in Codex Atlanticus by Leonardo da Vinci, c. 1504. © Veneranda Biblioteca Ambrosiana – Milano/ De Agostini Picture Library. 177

Figure 58 / Plate 9. *Still Life with Kettle*, c. 1869 (oil on canvas), by Paul Cézanne. Musée d'Orsay, Paris, France. Photo Credit: Réunion des Musée Nationaux/ Art Resource, NY. Eye tracking courtesy of Ellen Poliakoff and Andrew Bangham. 179

Figure 62 / Plate 2. *Portrait of Ambroise Vollard* (1868–1939), 1909 (oil on canvas), by Pablo Picasso (1881–1973). Pushkin Museum, Moscow, Russia/ Giraudon/ The Bridgeman Art Library. © 2011 Estate of Pablo Picasso/ Artists Rights Society (ARS), New York. 201

Figure 63. *My Wife and My Mother-in-Law*. From Boring 1930. 203

Figure 70. Some of the areas of the brain involved in visual processing. Adapted from *Scientific American,* Nov. 1999, page 48. 215

Figure 72. *Portrait of Maria Trip*, 1639, by Rembrandt van Rijn. Rijksmuseum, Amsterdam. *Portrait of a Girl*, 1917–18, by Amedeo Modigliani. Private collection. 221

Figure 75. *Portrait of Jeanne Hébuterne*, 1919, by Amedeo Modigliani. Art & Fragrances S.A. Cham/Switzerland. *Oscar Miestchaninoff*, 1917, by Amedeo Modigliani. Private collection. 225

Figure 78. Ocular dominance patterns in primary visual cortex of macaque monkey. From the D. H. Hubel and T. N. Wiesel, Ferrier Lecture (1977). *Proc R Soc Lond B* 198: 1–59, fig. 24(a). 235

Figure 79 / Plate 11. *St Jerome and the Lion in the Monastery*, 1501–9 (oil on canvas), by Vittore Carpaccio (c.1460/5–1523/6). Scuola di San Giorgio degli Schiavoni, Venice, Italy/ Giraudon/ The Bridgeman Art Library. 238

Figure 80 / Plate 12. *Hercules and the Hydra,* c. 1475 (tempera on panel), by Antonio Pollaiolo (1432/3–1498). Galleria degli Uffizi, Florence, Italy/ Alinari/ The Bridgeman Art Library. 247

Figure 81 / Plate 13. *Tobias and the Angel,* 1470–80, by Andrea del Verrocchio (1436–1488) and Leonardo da Vinci (1452–1519). National Gallery, London. © National Gallery, London/ Art Resource, NY. *Tobias and the Angel* (tempera on panel), by Antonio Pollaiuolo and Piero Pollaiuolo. Galleria Sabauda,

Turin, Italy/ Alinari/ The Bridgeman Art Library. 248

Figure 83 / Plate 14. *Still Life with Quince, Apples, and Pears,* c. 1885–87 (oil on canvas), by Paul Cézanne (1839–1906). Bequest of Charles A. Loeser, 1952. White House Historical Association (White House Collection): 523. 276

Figure 84 / Plate 15. *Apples and Pears*, c. 1885–90, by Pierre-Auguste Renoir (1841–1919). Musée de l'Orangerie, Paris. Jean Walter and Paul Guillaume collection. Photo Credit: Réunion des Musée Nationaux / Art Resource, NY. 277

Figure 85. *Print Gallery,* 1956. M.C. Escher's "Print Gallery" © The M.C. Escher Company-Holland. All rights reserved. www.mcescher.com. 279

Index

action-learning loops, 187–88
action potentials, 132–35
adaptation: in evolutionary contexts, 31, 40, 56–58, 125–26; as instinctive responses of single-cell organisms, 124–27, 135
the "Age of Bacteria," 118, 288
algae, 115–16, 118, 124, 125*f*, 278
ambiguity, 202, 217–18
Anatomy of the Horse (Stubbs), 105, 106*f*, 287
Anna Karenina (Tolstoy), 172
Antirrhinum, 101–2, 287
Aplysia californica. *See* sea slugs
apoptosis, 104
apples: in learning analogies, 155–63, 169–71, 178–80, 194–96, 219, 241; in linguistic analogies, 55–56; origins and evolution of, 13, 14, 18–20, 28, 30–31, 33, 58, 284; in paintings, 97–101, 275–76, PLATES 5, 6; predictable features of, 212, 239–41; in rainbow world analogies, 272; in regulatory protein analogies, 72–73, 76–80, 85–87, 92; in science, 147, 200, 274–76; sensory-motor responses to, 143, 146–47, 192, 196, 198, 270–72; Turing's poisoning of, 61
Apples and Biscuits (Cézanne), 97–99, 100*f*, PLATES 5, 6
Apples and Pears (Renoir), 275, 277*f*, PLATE 15
Arabidopsis thaliana, 44, 130, 288
Arcimboldo, Giuseppe: "The Four Seasons": *Spring, Summer, Autumn, Winter*, 127, 129*f*, 202, PLATE 8; *The Librarian*, 127, 128*f*, 202, PLATE 7
The Art of Genes (Coen), 10, 283
association, 153–54
Autumn (Arcimboldo), 127, 129*f*, 202, PLATE 8
averaged portraits, 220–28, 294–95, PLATE 10
axons, 136–37

Ball, Philip, 284
ballistic movements, 178–81
Bangham, Andrew, 295
Bard, Jonathan, 94, 287
Barrow, Isaac, 250
Barto, Andrew, 157, 290
Barton, N. H., 284
Bernard, Émile, 275, 298
Berns, G., 298
Bicoid gradients, 76–80, 92–93, 286, 287
Blackmore, S., 298
bladder wrack (*Fucus vesiculosus*), 118–20, 278
bombards, 176, 177*f*, 189, 291
Bonner, John Tyler, 285, 288
Boring, Edward, 203*f*
Botticelli, Sandro, 246, 248
bottom-up processing, 216–17
Bray, Dennis, 288
Breugel the Elder, Pieter, 53, 54*f*
Brown, David, 249, 296
Bruce, V., 293, 295
Burdon-Sanderson, John, 132
Byron, George Gordon, Lord, 256–57, 297

calibration, 176–78; in action-learning loops, 187–88; of ballistic movements (jumping eyes), 178–84, 291–92; of smooth movements, 188–93, 292
capturing relationships, 126–40, 278; in art, 94, 200, 220, 275; in development, 116, 118–22, 140, 278, 280; in evolution, 51, 58, 118–21, 278, 280; in learning, 1, 172, 190, 216, 280
caricature, 226, 295
Carpaccio, Vittore, 237, 238*f*, PLATE 11
Carpenter, Roger H. S., 292
Carroll, Sean, 286, 287, 288
carving, 116, 140, 147, 219, 243, 278, 280
categorization, 147–49
cave paintings of Lascaux, 20, 21*f*, 251
cell death, 103–4
cerebral cortex, 107–8, 287

Cézanne, Paul, 90, 228–29, 251; *Apples and Biscuits*, 97–99, 100*f*, PLATES 5, 6; *The Garden at Les Lauves*, 84, 85*f*, PLATE 4; obsessions of, 275, 298; *Portrait of Ambroise Vollard*, 34, 35*f*, 284, PLATE 1; reputation of, 252; *Still Life with Kettle*, 178, 179*f*, PLATE 9; *Still Life with Quince, Apples, and Pears*, 275, 276*f*, PLATE 14
change, 122–27
chatrang/chaturanga, 3–4
chess, 3–4, 6–7, 283
chicken and egg problem, 93
Chlamydomonas. *See* green algae
chromosomes, 19, 21–22. *See also* DNA
classical conditioning, 163–64, 290
Coen, Enrico, 10, 283, 285, 287
cognition, 175–76. *See also* learning; vision
color, 271–72
combinatorial richness, 16; in cultural change, 257–60, 264, 268, 297; in development, 88–90, 109, 115; in evolution, 40–44, 56, 59–60, 284; in learning, 168–69, 192–93, 198, 220; in painting, 48, 258
communication, 263, 267, 269, 297
competition: chance variation and fixation in, 27–28, 30; in cultural change, 254–55, 263–64, 267–68, 296, 297; definition of, 67; in development, 65–67, 80–83, 98–104, 115; in evolution, 16, 24–33, 39–40, 58–60; genetic drift in, 28–31; interaction with reinforcement of, 29–33, 58–60, 66, 67*f*, 80–83, 98–104, 167, 219–20, 236, 254, 263–64; in interpretation of perception, 217–20; in learning, 167, 168, 174, 186, 188, 213, 217–20, 236; linkage with cooperation of, 39–40, 256–57, 268, 296
Conditioned Reflexes and Neuron Organisation (Kornorski), 155, 289
conditioning, 153–55; classical conditioning as, 163–64, 290; dopamine-releasing neurons in, 155–57; operant (instrumental) conditioning as, 163–64, 178, 290, 293; punishment in, 164, 165; secondary conditioning as, 170–71, 290; Temporal-Difference Learning (TD-learning) in, 157–64, 168, 169–73, 268–69, 290, 291
consequences, 171
context, 50; in cultural change, 248–50, 256, 261, 296; in development, 92–93, 109; in evolution, 50–52, 59; in learning, 165
Cook, Michael, 240, 296
cooperation: communication/interaction with others in, 197–98, 256, 263, 297; in cultural change, 255–57, 264, 267–68, 296, 297; in development, 87–88, 115; in evolution, 36–40, 48, 50, 59–60, 284; interaction with competition of, 39–40, 256–57, 268, 296; in learning, 168, 174, 188, 197–98, 220; role of proximity in, 38–39, 48, 50, 88, 109, 116–17, 168–69
cooperative binding, 286
copying (replication): in cultural change, 251–52; in evolution/of DNA, 20, 21–23, 32. *See also* persistence
Corner, E. H., 288
correlation neurons, 210–11, 213, 222–28, PLATE 10
Coyne, Jerry, 264, 298
creativity, 143–47, 228–31, 249–50, 293. *See also* cultural change; painting
Credi, Lorenzo di, 246
Crick, Francis, 283
cultural change, 1–3, 11, 228–31, 244–48, 263–65; combinatorial richness in, 257–60, 264, 297; communication in, 263, 267, 297; competition in, 254–55, 263–64, 267–68, 297; cooperation in, 255–57, 264, 267–68, 297; creative acts in, 228–31, 249–50, 293; economic contexts of, 256; feedback loops in, 254–55, 263–64, 296; framing of transformations in, 269–80; historical place of, 5, 259–60, 267–70, 297; memes in, 264, 297–98; persistence in, 250–52, 260–61, 264, 267; population variation in, 250, 252, 263–64, 267; recurrence in, 261–63, 264, 268–69, 298; reinforcement in, 252–54, 263–64, 267; social context of, 248–50, 261, 296; speed in, 263, 297; values in, 252–54, 262–63, 267, 298
cultural space, 258–63, 264, 268, 280, 297
curiosity, 193–94

Darwin, Charles, 13, 269; on competition, 24–25, 284; education of, 250; *HMS Beagle* voyage of, 24; on long-term memory, 165–66, 290; publishing with Wallace of, 255; on scientific creativity, 230, 295; on sea slugs, 135–36, 289; on sexual reproduction, 52–53; on the Venus fly trap, 132, 288. *See also* evolution
Darwinian selection. *See* evolution
Dawkins, Richard, 284, 297–98
Dayan, Peter, 157, 290, 292, 293
Dean, Bashford, 111–13, 120, 287
decision-making, 196, 293
deformation, 97–104, 287; in three dimensions, 105–8; through death, 102–4; through growth, 97–102; through rearrangement, 104
dendrites, 136–37
Dennet, Daniel, 284
development, 1–3, 11, 108–21; adaptation in, 124–27, 135; of an embryo, 84–110; cell death in, 103–4; cell specialization in, 117; combinatorial richness in, 88–90, 109, 115; competition in, 65–67, 80–83, 98–104, 115; cooperation in, 87–88, 115, 116–17; cultural framework of, 269–80; cyclical character of, 110; deformation in, 97–104, 287; epigenetic changes in, 286; feedback loops in, 66, 67*f*, 69–70, 93, 108–9; in fruit flies, 73–80, 85–87, 88, 286; gradients and threshold levels in, 76–80, 92–93, 286; growth in, 93–97, 116; historical place of, 4–5, 114–15, 233–37; levels of organization in, 114–21, 122, 287–88; patterning at the cellular level in, 68–70; patterning of multicellular organisms in, 70, 71*f*, 88; persistence in, 65, 80, 81–82, 114–15; population variation in, 64–65, 80, 81–82, 114; recurrence in, 90–93; reinforcement in, 65–66, 67*f*, 80–83, 98–104, 114–15; relationship with evolution of, 120–21, 288; role of proximity in, 88, 109, 116–17; shifting contexts in, 92–93, 109; three-dimensional structures in, 105–8, 287; transition to learning from, 233–37; Turing's work in, 61–70, 80, 236
developmental space, 89–90, 168–69, 258, 278
Diamond, Jared, 297
Dictyostelium, 104, 104*f*
diffusion, 114
directness, 242
discrepancy neurons, 157–63, 169–74, 268–69, 291. *See also* Temporal-Difference Learning (TD-learning)
DNA: chromosome pairs of, 19; coding for protein in, 37–39; combinatorial richness in, 41–48; epigenetic changes in, 286; heritable variation and mutation in, 18–19, 22; molecular cohesion of, 20–21; recombination of, 36–37, 38; replication of, 21–23, 32. *See also* genes
dogs (of Pavlov), 152–55
dopamine, 155–57, 161–63, 169–70, 290
Drosophila melanogaster. *See* fruit flies
drug addiction, 170
Dürer, Albrecht, 122, 123*f*, 127

E. coli. See *Escherichia coli*
Eagleman, David, 292, 295, 296
ecosystems, 57–58, 285
Edelman, Gerald, 2, 7, 283
efference copy, 184
Einstein, Albert, 272–73, 275, 298
embryonic development. *See* development
epigenetic changes, 286
Escher, M. C., 279*f*, 280
Escherichia coli, 68–70; instinctive responses in, 237, 242–43; movement strategies of, 125–26, 288; response to lactose in, 115, 124, 132, 147
Essay on the Principle of Population (Malthus), 24–25
evolution, 1–3, 10–11, 58–60; adaptation and extinction in, 31, 40, 56–58, 125–26; branching tree diagrams of, 111–13; combinatorial richness in, 40–48, 56, 59–60; competition in, 16, 24–33, 39–40, 58–60; cooperation in, 36–40, 48, 50, 59–60, 284; creation of niches in, 57–58, 285; cultural framework of, 269–80; discovery of, 6; feedback loops in, 23, 59–60, 285;

evolution (*continued*)
historical place of, 4, 113, 233; instinct in, 237–43, 296; open-endedness of, 110; persistence in, 15, 20–23, 32, 59–60; population variation in, 16–20, 22, 32, 59–60, 81; recurrence in, 48–56; reinforcement in, 15, 16, 23–24, 29–33, 58–60, 82; relationship with development of, 120–21, 288; role in learning of, 237–42; role of proximity in, 38–39, 48, 50; sexual reproduction in, 52–53, 56, 237–39; shifting contexts in, 50–52, 59; speciation in, 52–58, 60, 284; traditional accounts of, 13–14, 284; visualization of, 45–48
expectations. *See* predictive learning
exploratory learning, 193–94
extinction, 58, 285
eye-movement neurons, 181–86
eyes. *See* vision

fear, 239–40
feedback loops, 23; in cultural change, 254–55, 263–64, 296; in development, 66, 67*f*, 69–70, 93, 108–9; in evolution, 59–60, 285; in learning, 167, 174, 186, 188, 197–98, 219–20, 236
Fitch, W. T., 297
fitness, 29–33, 58–59
fixation (in a population), 27–28, 30
flagella, 124–25, 278
flexibility, 242–43, 296
flight, 47, 51
Florentine Renaissance, 248–50, 251, 254, 256–57, 296. *See also* Leonardo da Vinci; Michelangelo
"The Four Seasons": *Spring, Summer, Autumn, Winter* (Arcimboldo), 127, 129*f*, 202, PLATE 8
frameworks, 274–80
framing, 3, 232–43, 269–70, 278–80
Freud, Lucien, 295
Frith, Chris, 198, 293
frontal eye field, 292
fruit flies (*Drosophila melanogaster*), 73–80, 85–87; Bicoid and Hunchback proteins in, 76–80, 92–93, 286, 287; cells with multiple nuclei in, 77, 88; neuroblasts in, 73–74, 286; overlapping pattern frameworks in, 90–93, 95; signaling and receptor system in, 74–76. *See also* development
Fucus vesiculosus, 118–20, 278
fusiform face area, 202–3, 293

Gallese, Vittorio, 197, 293
ganglia, 137–39
gap genes, 87
The Garden at Les Lauves (Cézanne), 84, 85*f*, PLATE 4
Gavrilets, Sergey, 285
Gayford, Martin, 295
Geffroy, Gustave, 275, 298
general meaning (of principles), 15–16
genes: regulatory regions of, 72–73; RNA coding of, 71–72, 73; switching mechanisms of, 70–73. *See also* DNA
genetically-based disease, 36
genetic drift, 28–31
genetic space, 44–48, 89, 168–69, 258, 278
genomes: chromosomes pairs in, 19; combinatorial richness in, 44–48, 88–90, 284; heritable variation and mutation in, 18–19, 22
Ghirlandaio, Domenico, 246
gill-withdrawal reflex, 137–39
Gladwell, M., 296
glutamate, 138–39
God Creating Adam (Michelangelo), 47, 48*f*
Gould, Steven Jay, 118, 283, 284, 288
gradients, 76–80, 92–93
gravity, 272–73
green algae (*Chlamydomonas*), 124, 125*f*, 278
Greene, Brian, 298
Gregory, Richard, 190, 292
growth, 93–104
Gutenberg, Johannes, 251, 296

habituation, 134, 141, 142*f*
Haeckel, Ernst, 2, 7, 283
Haldane, J.B.S., 298
Hanna, Andy, 295
hatless-long-head neuron, 223–28, PLATE 10

Hawkins, Jeff, 289
Haykin, S., 291, 294, 295
Hebb, Donald, 155, 289, 294
Hebbian neural networks, 294, 295
Heidegger, Martin, 291
helmet evolution (Dean), 111–13, 120
Helmholtz, Hermann von, 196–97, 293
helpfulness, 240–41
Henslow, John, 250
Hercules and the Hydra (Pollaiolo), 246, 247*f*, PLATE 12
Hinton, Geoff, 296
HMS Beagle, 24
homeostatic mechanisms, 124–26
Homer, 166, 290
Homo habilis, 259
Homo sapiens, 259–60, 297
Howard, J., 284
Hox genes, 90
Hubbell, Stephen, 285
Hubel, David, 234, 236, 295–96
human creativity. *See* creativity
human genome, 44
human nervous system. *See* nervous system
Hunchback proteins, 76–80

identifying commonalities, 14–16
incentives, 195–96, 292–93
individual variation, 17–19, 59, 64
instinct, 237–42, 296
instrumental conditioning, 163–64, 178, 290, 293
interpretation of perception, 200–204; ambiguity in, 202, 217–18; averaging of complex images, 220–28, 294–95, PLATE 10; correlation neurons in, 210–11, 213, 222–28; in creative acts, 228–31; facial recognition in, 202–3, 293; integration of, 215–16; invariant representations in, 212, 294; model formation in, 211–13; multiple levels of, 213–16; the neural eye in, 204–11, 220–28, 294–95; resolution of competition in, 217–20; selection neurons in, 204–5; two-way interactions in, 216–17
invariant representations, 212, 294

Jerome, Saint. *See* St. Jerome
Johannsen, Wilhelm, 258
Johnson, Mark, 293
jumping eyes, 178–81

Kamin, Leon, 154, 289
Kandel, Eric, 136, 289
Kant, Immanuel, 298
Kennaway, J. R., 287
knee reflex, 143
Knoll, Andrew, 283, 285
Koch, C., 294
Konorski, Jerzy, 155, 289, 290, 291

Labours of Hercules (Pollaiuolo), 246
Lakoff, George, 293
Lane, Nick, 283, 285
language evolution and development, 297
Large Piece of Turf (Dürer), 122, 123*f*, 127
Lascaux cave paintings, 20, 21*f*, 251
lateral geniculate nucleus (LGN), 215
law of compound interest, 23–25, 29
law of mass action, 64–65
learning, 1–3, 11, 168–74, 228–31, 242–43; action-learning loops in, 187–88; adaptation as, 124–27; calibration of actions in, 176–84, 187–99, 291–92; categorization in, 147–49; combinatorial richness in, 168–69, 192–93, 198, 220; competition in, 167, 168, 174, 186, 188, 213, 217–20, 236; conditioning of associations in, 153–57, 163–65, 170–71, 178, 290, 293; connecting actions with goals in, 176–78, 193–96; cooperation in, 168, 174, 188, 197–98, 220; creative acts in, 228–31, 249–50, 293; cultural framework of, 269–80; decision-making in, 196, 293; expectations and discrepancies in, 157–64, 168, 169–74, 196, 268–69, 291; feedback loops in, 167, 174, 186, 188, 197–98, 219–20, 236; habituation in, 134, 141, 142*f*; historical place of, 5, 233–37; imitation in, 197–99, 293; incentives in, 195–96, 292–93; instinct and values in, 237–43, 267, 296; interpretation of perception in, 200–231; of new

learning (*continued*)
relationships, 149; persistence in, 132–33, 165–66, 168, 186, 188, 213, 220; population variation in, 166, 168, 174, 186, 188, 213, 220; predictive learning in, 150–64, 289; punishment in, 164, 165; recurrence in, 173–74, 192–93, 198, 220; reflex responses in, 136–40; reinforcement in, 166–67, 168, 174, 186, 188, 213, 219–20, 236; response to environmental change in, 127–35, 147–49, 194–95, 239, 242–43, 288–89, 296; role of proximity in, 168–69, 188; role of time and space in, 122–27, 134–35, 140–43, 288–89; self-initiated exploration in, 193–94; sensitization in, 142–43; sensory stimulation and integration in, 143–47, 194–97; shifting contexts in, 165; temporal organization in, 153–55; transition from development to, 233–37; visual stimulation in, 175–99. *See also* nervous system; spatial organization; Temporal-Difference Learning; temporal organization; vision
Leonardo da Vinci, 291, 295; apprenticeship of, 245–48, 249, 251, 254–58; competition with Michelangelo of, 297; on cooperative work, 257; library of, 251; military inventions of, 176, 177*f*, 189, 253, 255, 261, 291; on painting, 256, 261–62, 297; scientific values of, 262; social context of, 248–50; *Tobias and the Angel*, 246, 248*f*, 254, 255–56, PLATE 13
Leroi, Armand, 286
Levi-Montalcini, Rita, 287
The Librarian (Arcimboldo), 127, 128*f*, 202, PLATE 7
Libyan Sibyl (Michelangelo), 150, 151*f*
life's creative recipe, 60, 267–70; human self-framing of, 270–80; identifying commonalities in, 14–16; seven principles of, 8–12. *See also* cultural change; development; evolution; learning; specific principles, e.g. persistence
linguistic diversity, 53–55, 284–85
long-term habituation, 141, 142*f*
long-term sensitization, 143
long-thin-head neuron, 223–28, PLATE 10
MacGregor, Neil, 297
Mairan, Jean-Jacques d'Ortous de, 128–29
Malthus, Thomas, 24–25
Malus pumila, 13, 284
The Man Who Mistook His Wife for a Hat (Sacks), 293
Mayr, Ernest, 284
McNeill, J. R., 245, 296
McNeill, William, 245, 296
mechanisms of transformation, 1–3; historical relationships of, 4–5; human self-framing of, 270–80; identifying commonalities in, 14–16; relationships of form in, 5–8; seven principles of, 8–12
Meinhardt, Hans, 285
memes/memetics, 264, 297–98
Michelangelo, 248–50, 296; competition with Leonardo of, 297; *God Creating Adam*, 47, 48*f*; innovations in painting of, 51–52; *Libyan Sibyl*, 150, 151*f*
military inventions, 176, 177*f*, 189, 253, 255, 261, 291
mimosas, 128–29, 134, 289
Mineka, Susan, 240, 296
Min proteins, 68–70, 286
mirror neurons, 197–98, 293
model formation, 211–13
Modigliani, Amedeo, 220–28, 295; *Oscar Miestchaninoff*, 224, 225*f*, 227–28; *Portrait of a Girl*, 220, 221*f*; *Portrait of Jeanne Hébuterne*, 223–24, 225*f*
molecular cohesion, 20–21
Monet, Claude: innovations in painting of, 51; *The Rue Montorgueil*, 47–48, 49*f*, PLATE 3
Montague, Read, 157, 290, 293
motor area, 145*f*, 146–47
motor neurons, 138
Mu Qi, 8, 9*f*
mutation: in coding of regulatory proteins, 92; in evolution, 18–19, 22, 82, 286; principle of cooperation in, 36–37
My Wife and My Mother-in-Law (Boring), 203*f*

natural selection. *See* evolution
Neisser, Ulrich, 292
nervous system, 143–47; expectation and discrepancy learning in, 157–64, 268–69,

291; fusiform face area, 202–3, 293; mirror neurons in, 197–98, 293; predictive operation of, 155–63; presynaptic inhibition in, 204–5, 293–94; primary motor area of, 145*f*, 146–47; primary somatosensory area of, 143–46, 289; reflexes in, 143; structural plasticity of, 290; visual areas of, 187, 214–16, 233–34, 291–92, 295. *See also* learning
neural Darwinism, 2, 7, 283
neural eye movement, 204–8, 220–28, 294–95
neural space, 169, 174, 258, 279–80
neural tubes, 107–8
neuroblast development, 73–74, 286
neurons, 136–37, 155–57
neurotransmitters, 137–39; dopamine, 155–57, 161–63, 169–70, 290; glutamate, 138; presynaptic inhibition of, 204–5, 293–94
neurotrophins, 103–4, 287
Newton, Isaac, 250, 272–73, 275
niches, 57–58, 285
Noble, Denis, 284
Nüsslein-Volhard, C., 286, 287

ocular dominance patterns, 234–36, 295–96
The Odyssey (Homer), 166
One Thousand and One Nights, 232–33, 241, 270
operant conditioning, 163–64, 178, 290, 293
The Organisation of Behavior (Hebb), 155, 289
Oscar Miestchaninoff (Modigliani), 224, 225*f*, 228

painting, 260–61; averaging of portraits in, 220–28, 294–95; Byron on, 256; caricature in, 226, 295; combinatorial richness in, 48, 258; innovations in, 51–52; interplay of color and space in, 34, 35*f*, 84, 90, 284, PLATE 1; interplay of expectations and discrepancies in, 228–29, 295; Leonardo on, 256, 257, 261–62, 297; portrayal of movement in, 47–48, 49*f*; refining brushwork in, 94–96; self-portraiture in, 280; values in, 262; Xie He's ingredients for quality in, 8, 9*f*
patterning, 73–76, 80–83, 87, 271; gradients and threshold levels in, 76–80, 92–93, 286; interaction with growth of, 94–96; of multicellular organisms, 70, 71*f*, 88; in ocular dominance, 234–36, 295–96; overlapping of patterns in, 90–93; polarization in, 75–76; of single-celled organisms, 68–70; switching mechanisms in, 70–73; Turing's principles of, 61–70, 80, 236
Pavlov, Ivan Petrovich: on punishment, 164; salivating dog of, 152–55, 163, 170–71, 289
Penfield, Wilder, 143–47, 289
Persimmons (Mu Qi), 8, 9*f*
persistence: communication in, 267; in cultural change, 250–52, 260–61, 264, 267; in development, 65, 80, 81–82, 114–15; in evolution, 15, 20–23, 32, 59–60; in learning, 132–33, 165–66, 168, 186, 188, 213, 220; molecular cohesion in, 20–21
person models, 213
pheromones, 114
Picasso, Pablo, 200, 201*f*, PLATE 2
Pinker, Steven, 285
plankton, 115–16
poetry, 229
Poggio, T., 295
polarization, 75–76, 286
Pollaiuolo, Antonio, 245–48, 296; *Hercules and the Hydra*, 246, 247*f*, PLATE 12; *Tobias and the Angel*, 246, 248*f*, 251, 254, 255, PLATE 13
Pollaiuolo, Piero, 246, 248*f*, 251, 254, 255, PLATE 13
population variation: in cultural change, 250, 252, 263–64, 267; in development, 64–65, 80, 81–82, 114; in evolution, 16–20, 32, 59–60, 81; heritable variation and mutation in, 18–19, 22; in learning, 166, 168, 174, 186, 188, 213, 220; sexual reproduction in, 19
Portrait of a Girl (Modigliani), 220, 221*f*
Portrait of Ambroise Vollard (Cézanne), 34, 35*f*, 200, 284, PLATE 1
Portrait of Ambroise Vollard (Picasso), 200, 201*f*, PLATE 2
Portrait of Jeanne Hébuterne (Modigliani), 223–24, 225*f*
Portrait of Maria Trip (Rembrandt), 220, 221*f*
Possible Worlds (Haldane), 298
Powers, William, 292

predictive learning, 150–57, 289; expectations and discrepancies in, 157–63, 168, 169–74, 196, 268–69, 290, 291; relational aspects of, 170–73; shifting contexts in, 165; temporal organization in, 153–55. *See also* learning
presynaptic inhibition, 204–5, 293–94
primary motor area, 145*f*, 146–47
primary somatosensory area, 143–46, 289
primary visual cortex, 214–16, 233–34
principal component analysis, 295
principles. *See* combinatorial richness; competition; cooperation; persistence; population variation; recurrence; reinforcement
Print Gallery (Escher), 279*f*, 280
printing press, 251, 296
probability, 18–19
prosopagnosia, 202–3, 293
protein codes, 37–39, 44–48, 284
Proust, Marcel, 141
proximity: in development, 88, 109, 116–17; in evolution, 38–39, 48, 50; in learning, 168–69, 188
Ptashne, Mark, 288

quantum mechanics, 273–75, 276, 298

rainbow world, 271–72, 298
Ramachandran, V. S., 293, 297
Ramón y Cajal, Santiago, 154–55, 289
randomness, 17
The Rational Optimist (Ridley), 297
recognition, 228
recombination, 19, 36
recurrence, 16; in cultural change, 261–63, 264, 268–69, 298; in development, 90–93; in evolution, 48–56, 59–60; in learning, 173–74, 192–93, 198, 220
Red Queen Effect, 285
Rees, Martin, 273, 298
reflex responses, 136–40
regulatory proteins, 72–73; in *Arabidopsis*'s monitoring of light, 130, 288; of the cerebral cortex, 107–8, 287; combinatorial richness in, 88–90; control of growth and cell death by, 101–4; cooperation in, 87–88; recurrence in, 90–92; of single-celled organisms, 114–17, 124
Reichardt, W., 292
reinforcement: communication in, 267; in cultural change, 252–54, 263–64, 267; in development, 65–66, 67*f*, 80–83, 98–104, 115; in evolution, 15, 16, 23–24, 29–33, 58–60, 82; interaction with competition of, 29–33, 58–60, 66, 67*f*, 80–83, 98–104, 167, 219–20, 236, 254, 263–64; law of compound interest in, 23–24, 29; in learning, 166–67, 168, 174, 186, 188, 213, 219–20, 236
relationships. *See* capturing relationships
relativity, 273–75, 298
Rembrandt van Rijn, 220–28, 295
Renaissance: Florentine artists of, 248–50, 251, 254, 256–57, 296; scientific achievements of, 262
Renoir, Pierre-Auguste, 275, 277*f*, PLATE 15
replication. *See* copying
Reynolds, Joshua, 251, 258, 296
Ridley, Matt, 285, 297
Rizzolatti, Giacomo, 197, 293
RNA, 71–72, 73
Romo, Ranulfo, 155–57, 161, 193–94, 290
Rubisco (ribulose biphosphate carboxylase) proteins, 45–46, 51, 119–20, 278
The Rue Montorgueil, Paris (Monet), 47–48, 49*f*, PLATE 3
Rutherford, Ernest, 257, 297

saccades, 178–81, 184–87, 291
saccadic suppression, 291
Sacks, Oliver, 40, 284, 293
scales: in art, 94; in cultural change, 257; in development, 92–97, 109, 117–21, 135, 147; in diffusion, 64; in evolution, 117–21, 242, 269; in human cooperation, 257; in instinctive responses, 126, 135, 140, 147–48; in nervous systems, 147, 195, 213–16, 220, 234; in the physical world, 40, 55, 147, 273–74; in populations, 17; in time, 125, 132, 189, 195, 242, 269
Scheherazade, 232–33
Schultz, Wolfram, 155–57, 161, 163, 193–94, 290

science, 3, 295; core values in, 262–63; creativity in, 229–31; cultural frameworks of, 269–80
sea slugs (*Aplysia californica*), 135–40, 289; central nervous system of, 139–40; gill-withdrawal reflex of, 137–39, 148–49, 237; habituation in, 141, 142*f*
seaweed, 118–20
secondary conditioning, 170–71, 290
segment polarity genes, 90
Sejnowski, Terry, 157, 290, 294
selection neurons, 204–5
self-carving. *See* carving
self-portraiture, 280
self-promotion. *See* reinforcement
sensitization, 142–43
sensory integration, 143–47
sensory neurons, 137–40
sexual reproduction, 52–53, 56; chromosomes in, 19; gestation time in, 118; instinct in, 237–39, 242; learning in, 242–43; pheromones in, 114; recombination in, 19
Sforza, Ludovico, 253
Sherrington, Charles, 155, 289
short-term habituation, 141, 142*f*
short-term sensitization, 143
Sinbad the Sailor, 232–33, 241, 270
single-celled organisms: adaptation in, 124–26, 135; multiple nuclei in, 116; organization of development in, 114–21, 287–88; patterning in, 68–70; reproduction of, 118
Sistine Chapel, 250; *God Creating Adam*, 47, 48*f*; *Libyan Sibyl*, 150, 151*f*
Skinner, B. F., 290
slime mold (*Dictyostelium*), 104, 104*f*
Smith, J. M., 284
snapdragons (*Antirrhinum*), 101–2, 287
somatosensory area, 143–46, 289
somites, 105, 107*f*
spatial organization, 114–21, 287–88; of animal nervous systems, 140; of Venus fly trap motion, 133. *See also* proximity
special relativity, 273–75, 298
speciation, 52–58, 60, 284, 285; creation of niches in, 57–58, 285; extinction in, 58
specific meaning (of principles), 15–16
Spirov, A., 287
Spring (Arcimboldo), 127, 129*f*, 202, PLATE 8
St. Jerome, 237, 238*f*, 240–41, PLATE 11
St. Jerome and the Lion in the Monastery (Carpaccio), 237, 238*f*, 240–41, PLATE 11
statistical reasoning: chemical reaction rates in, 64–65, 66; diffusion in, 64; individual variation in, 17–19, 64; law of mass action in, 65; population behavior in, 17, 19–20, 64; probability in, 18–19
Stern, David, 288
Still Life with Kettle (Cézanne), 178, 179*f*, PLATE 9
Still Life with Quince, Apples, and Pears (Cézanne), 275, 276*f*, PLATE 14
storytelling, 172, 266
Stratton, George M., 190–91, 292
structural plasticity, 290
Stubbs, George, 93–94, 287; *Anatomy of the Horse*, 105, 106*f*, 108; *Zebra*, 95*f*
style, 220–28
Summer (Arcimboldo), 127, 129*f*, 202, PLATE 8
superior colliculus, 292
Sutton, Richard, 157, 290
synapses, 136–37, 155–57
Szathmáry, E., 284

Temporal-Difference Learning (TD-learning), 157–64, 168, 169–73, 268–69, 290, 291
temporal organization, 122–24; of animal nervous systems, 140–43; habituation in, 141, 142*f*; in plants' responses to the environment, 130–35, 239; in predictions and expectations, 153–55; sensitization in, 142–43
three-dimensional deformations, 105–8
Tickle, Cheryll, 286
Tobias and the Angel (Pollaiuolo brothers), 246, 248*f*, 251, 254, 255, PLATE 13
Tobias and the Angel (Verrochio and Leonardo), 246, 248*f*, 254, 255–56, PLATE 13
Tolstoy, Leo, 172
top-down processing, 216–17
Toriyama, Hideo, 134, 289
touch, 143
The Tower of Babel (Breugel the Elder), 53, 54*f*

trade-offs, 56, 58, 118, 242
Turing, Alan, 61–70, 236, 269, 285, 296

unicellular organisms. *See* single-celled organisms
unified theories of living transformations, 2–3, 11

values: in cultural change, 252–54, 262–63, 267; in learning, 241–42, 267
Van Essen, David, 208, 294
van Eyck, Jan, 250
van Gogh, Vincent, 252
variation, 16–17, 252. *See also* population variation
Venus fly traps (*Dionaea muscipula*), 132–35, 147–48, 242, 278, 288
Verrocchio, Andrea del, 296; Leonardo's apprenticeship with, 245–49, 251; *Tobias and the Angel*, 246, 248*f*, 254, 255, PLATE 13
vertebrae, 105–7
Vespucci, Amerigo, 250
vision, 175–76, 271; action-learning loops in, 187–88; brain anatomy of, 187, 214–16, 233–37, 291–92; calibration of ballistic movements (saccades) in, 178–81, 184–87, 291–92; calibration of smooth movements in, 188–93, 292; facial recognition in, 202–3, 293; integration of, 215–16; interpretation of the neural eye in, 203–11, 220–28, 294–95; invariant representations in, 212, 294; neural interactions of, 181–84, 292; ocular dominance patterns in, 234–36, 295–96; Stratton's experiments in, 190–91. *See also* learning
visual form agnosia, 216, 294
visual-shift neurons, 183–85
visual-velocity neurons, 189–91
Vollard, Ambroise, 34, 35*f*, 200, 201*f*, 284, PLATES 1, 2
von der Malsburg, Christoph, 10, 283, 296

walking, 192
Wallace, Alfred Russel, 25, 255
war: analogies with chess of, 3–4, 6–7, 32, 83; helmet evolution for, 111–13, 120; Leonardo's inventions for, 176, 177*f*, 189, 253, 255, 261, 291
Wiesel, Torsten, 234, 236, 295–96
Winter (Arcimboldo), 127, 129*f*, 202, PLATE 8
Wolpert, Lewis, 286

Xie He, 8, 10

zebras, 94–96
Zebra (Stubbs), 95*f*
Zeki, Semir, 294